David Attenborough
Life Stories

PARADISEA MINOR

David Attenborough

Life Stories

Collins

BY THE SAME AUTHOR

Zoo Quest to Guiana (1956)

Zoo Quest for a Dragon (1957)

Zoo Quest to Paraguay (1959)

Quest in Paradise (1960)

Zoo Quest to Madagascar (1961)

Quest Under Capricorn (1963)

◇

The Tribal Eye (1976)

The First Eden (1987)

◇

Life on Earth (1979)

The Living Planet (1984)

The Trials of Life (1990)

The Private Life of Plants (1995)

The Life of Birds (1998)

The Life of Mammals (2002)

Life in the Undergrowth (2005)

Life in Cold Blood (2008)

Life on Air (new edition, 2009)

Published in 2009 by HarperCollinsPublishers Ltd.
77-85 Fulham Palace Road
London W6 8JB

Website: www.harpercollinspublishers.co.uk

ISBN 978 0 00 733883 2

Frontispiece: *A displaying bird of paradise in all its magnificence, drawn in 1873 by the great bird illustrator, Daniel Giraud Elliot.*

Printed in Great Britain by Butler Tanner & Dennis Ltd, Frome

Contents

Foreword

People sometimes ask me when and how I first got interested in animals. I have to admit that seems to me an extraordinary question. I find it hard to believe that any normal child is born without that interest. I recall taking a five-year old for a walk in a Devon meadow. He turned over a stone. 'Look,' he said, 'a slug. What a little treasure!' And quite right too. The way it moved, sliding along without the help of legs. Those eyes on the end of a couple of waving stalks that retracted when he touched them. Astonishing!

That wonder on first seeing a new kind of animal is only the first. The next question you ask is its name. I remember myself looking wide-eyed at an okapi in a zoo. What is it? I had to have a name for it. And once you have the name, there is the pleasure of trying to work out what its nearest relations are, what kind of animal it is. The okapi, I was told, is a cousin of the giraffe's but it lives in the forest where a giraffe's long neck would be not only unnecessary but extremely inconvenient. That made a very satisfying kind of sense. After that I read about how the world outside the Congo first became aware that there was such an animal. A British explorer travelling in the forests saw a man sitting on a wooden stool covered with a strange black and white striped skin. The skin, he was told came from a rare and very shy creature that lived in the depths of the forests. He had never heard of such an animal before. So he set off to find it.

We have all asked those kinds of questions during our lives. And humanity has been asking them across history as they discovered other continents for themselves and began to relish the fascination of the exotic.

Those questions are what this book is about. Its pages were first written as talks for radio. Here, happily, they have been amplified with illustrations, ancient and modern. They don't deal with the okapi. But they do describe other things just as strange, rare and wonderful.

David Attenborough

1

sloths

'What would you like to be,'
people sometimes ask,
'if you came back to earth
as an animal?'
The answer *I* give depends,
I suppose, on the company.

◁ Three-toed' is not really an accurate name because a sloth lacks true toes. All it
has, at the end of its feet, are curved claws. But those are all that it needs for a life
spent hanging by its legs. And sloths can close those claws downwards with such
strength that they can grip a branch very firmly indeed.

W hat would you like to be, people sometimes ask, if you came back to earth as an animal?

The answer *I* give depends, I suppose, on the company. If it is a bit racy, I dare say my mind would range over a number of mammals remarkable for the extravagance of their reproductive techniques.

In more sober company, I usually find a safer answer. 'A sloth', I say. And why? 'Because it spends most of its time hanging upside down from a branch in the tropical rain forest – dozing.'

The sloth is not only one of the laziest but one of the strangest of mammals. It spends most of its life asleep, but even when it is awake it only moves very slowly. No matter what the danger or the emergency, it is incapable of speed. There are two main kinds, defined by the number of toes that they have on their front legs. The two-toed sloth is relatively lively. It is also rather ferocious and will slash at you with its formidable curved claws. It is the three-toed which has really slothful habits and the one that interests me most. You are not likely to have seen it in zoos, for, unlike the two-toed which will eat all kinds of vegetables, the three-toed is really picky and will only accept leaves and fruits of a few particular kinds – which few zoos can supply. And in any case, it doesn't make an exactly dazzling exhibit since it spends most of its time in slumber.

It is about the size of a small, slim upside-down goat, if you can imagine such a thing. Its feet don't have soles but end in curved hooks – claws – from which it can hang without any muscular effort whatever. It is covered in a shaggy coat of coarse untidy hair, blotched in grey, cream and brown. The hair on its head is combed forward into a fringe. That on its body is parted down the middle of its belly with the hair running down its flanks towards its spine so that as it hangs upside down, the tropical rain, which it has to endure for several hours every day, does much the same.

The thick coat is itself a home for an abundant population of other creatures. There are ticks, a particular kind of beetle and moths of a species that occurs nowhere else. The hairs themselves have grooves down them in which grow green algae.

When the skins and skeletons of this extraordinary topsy-turvy beast reached Europe, artists given the task of illustrating the species in natural history encyclopaedias showed it standing upright on its four thin rod-like legs, even though the hooks on its feet and the fur on its body all pointed very obviously in what you might call the wrong way. After all, the artists must have said to themselves, it would be too absurd to show it as an animal that spent all its life upside down.

Because the sloth moves extremely slowly, there is little difficulty in catching one. But spotting one is not easy, because huddled in the crook of a branch or hanging beneath it, it looks like a bundle of dead leaves. The first one I ever encountered – in a forest in what was then British Guiana and is now Guyana – was pointed out to me by one of the forest-living Akawaio people in whose village we were staying. The tree wasn't difficult to climb – I was younger and nimbler in those days – and I was soon sitting astride a branch alongside it. The sloth looked at me amiably, from beneath its fringe with a distant but benign smile.

I was there making natural history films and I thought I would move the sloth to a branch where it would be easier to film. So I hauled up a liana that dangled nearby and had no problem in transferring the sloth's legs on to it, one by one, so that eventually I was able to lower it to the ground. We then took it back to the village and put it in an isolated tree the leaves of which were to the sloth's taste and which stood just outside our hut. There was little chance that it would stray because sloths, on the ground, are virtually helpless. Its legs are little more than hangers and only have slim ribbon-like muscles that are just about strong enough to enable it to transfer its grip from one branch to another, but quite incapable of supporting it standing upright. So on the ground, it lies on its stomach and waves its legs around with a vague sort of swimming motion and just about succeeds in crawling very slowly forward – which it clearly doesn't like because quite apart from the awkwardness of it all, it is very vulnerable to attack from ground predators such as jaguar.

The following morning, I went out to look at it as it hung from its branch and saw, to my delight, a tiny bedraggled baby sloth crawling very slowly from between its mother's hind legs and out on to her belly. One of the Akawaio women in the village told me a story about baby sloths. A pair of them – twins, I suppose – were sitting on their mother's stomach complaining that they were very hungry. Their mother craned her head forward and said 'Clamber on to that branch, children, and I will go off to find some leaves for you. But you must promise that you won't stray from this spot so that I will be able to find you again.' They promised – and away she went. A day passed. And then another. The babies got more and more hungry until eventually, on the fifth day, one of the babies whispered to the other 'Why don't we go and look for some leaves ourselves'. Whereupon their mother's voice came from the leaves nearby : 'I warned you,' she said. 'If you don't stay where you are, I won't go.'

In reality, there's not much whispering that goes on among sloths. They are only capable of a faint bronchial wheeze and an occasional whistle. These sounds are probably only made to deter a predator and certainly not to communicate with one another for sloths are virtually stone deaf. It is said that an enthusiastic zoologist, investigating the hearing ability of sloths, discharged his gun close to one, which merely slowly raised its head, blinked at the scientist – and then went back to sleep. Their eyesight is not much good either. In fact the only sense that seems at all acute is their sense of smell and it is that that they use to select their favourite leaves and fruit.

So how do these purblind half-deaf creatures find one another up in the branches when breeding time comes round? Well, it seems to be connected to their rate of digestion. That, like everything else, is extraordinarily slow. You might imagine that they would defecate with careless abandon, as they just hang there fifty feet up in the forest, and allow their dung to fall to the ground. But not a bit of it. When the time approaches, a sloth sets off on what may be quite a long journey to a special tree in its territory. It slowly clambers downwards until it reaches the ground where at the bottom of the tree, there is a big pile of sloth dung, to which all the sloths in the area have contributed. Sloths have a communal loo! And there it hangs within a foot or so of the ground beside the great smelly pyramid while it adds its own contribution.

The whole process is time-consuming and inherently, of course, dangerous. While a sloth is hanging beneath a branch it is very difficult to attack and dislodge – even for the harpy eagles that are its main predator. But beside its midden, any ground-living predator could get at it. So this behaviour must be of some considerable importance to the sloth. Perhaps the midden is the one and only conspicuous beacon in the sloth's dim and silent world – the only place where there is a reasonable chance of getting close to *another* sloth – and know that you are doing so. If that is so, then one would expect that sloth dung to be pungent. And I can tell you – it certainly is, even to the nostrils of a human being.

But what extraordinary quirk of evolution led sloths down into their dim and silent alley? Well, raw leaves – no matter what salad enthusiasts may tell you – are remarkably indigestible. Cows, deer and antelope deal with that, as every one knows, by giving a mouthful a good chew, sending it down to the stomach, letting it stew there for a bit and then bringing it back up again as cud, and chewing it a second time. And even after all that, the bulk of the meal, the cellulose, is so resistant to the digestive juices that it has no nutritive value and has to be ejected.

Out on the plains, an animal is very visible and an obvious target for meat-eaters – like lions. So plant eaters have to be either formidably – almost invincibly – big, like buffalo or elephant, or become such good runners, like antelope, that they can outpace most hunters.

But up in the trees and well-camouflaged, an animal can chew away and take all the time it wants to digest its meals. If that enables you to cope with particularly indigestible leaves that no one else wants, then you needn't remain particularly active or strong – or swift – and you can take all the time you like to digest them. In short, you become slothful.

There may be a moral somewhere in all this, but I am not sure what it is.

The artist who drew this illustration for a nineteenth century encyclopaedia must have realised, correctly, that a sloth's neck is particularly long . It has eight neck bones, which is even more than a giraffe's, and in some individuals even nine. Such a long neck, moreover, does give the animal the ability to swivel its head through 180° as it hangs upside down.

An Italian seventeenth-century collector, Cassiano del Pozzo, assembled a 'paper museum', several thousand drawings of animals and plants as well as strange stones and minerals and objects from antiquity. The artist who drew this sloth for him had clearly never seen the living animal and did his best to make sense of the skin and, perhaps, the skeleton with which he was presented. He shows that the animal's claws project directly from the long shaggy fur of its legs. Perhaps he did not dare to imagine that the animal spent its life upside down, though he must have been worried by the thought that if it walked with claws in such a position they would have worn down within days.

SLOTH Nº359

By the eighteenth century, when this engraving was published as an illustration to a survey of the world's mammals, naturalists knew the truth about the sloth's way of life. But although the artist portrays the animal's anatomy in a commendably accurate way, he still felt unable to show the sloth in its correct posture.

The two-toed sloth is very different from the three-toed. It is not merely that it has one toe fewer on each of its forelegs (though it has three on its hind). It is also considerably bigger, eats a greater variety of vegetation and is rather more aggressive. It may even be that the two-toed and three-toed sloths are descended from quite different ancestors and have become adapted to an upside-down life, each in its own separate way.

The hairy coat of the three-toed sloth is so long and shaggy, and the animal itself so inactive, that beetles and ticks as well as the moths shown here, have taken up residence in it. They did this so long ago that now they have all become species that occur nowhere else. Algae also lodge in the long grooves of the individual hairs so that the sloth's coat, at times, becomes visibly green.

The male three-toed sloth differs from the female in that it has a handsome and distinctive brown stripe fringed with gold running down the upper part of its spine. When – and indeed if – the female sloth notices this, we can only guess. ▷

2

Monstrous Flowers

I like flowers to have a certain
modesty. Shrinking violets are
more to my taste than
ostentatious rhododendrons. I
know, of course, that my
preferences are not shared by
most horticulturalists and
plant breeders.

◁ The titan arum flowering in the forest of southern Sumatra. The crimson spathe
has only been open for a day and within another day it will have closed up around
the central spike, the spadix.

I like flowers to have a certain modesty. Shrinking violets are more to my taste than ostentatious rhododendrons. I know, of course, that my preferences are not shared by most horticulturalists and plant breeders. They enjoy taking a shy wild species and with great skill producing a cultivated version that is ten times the species' original size. Wild cyclamen, for example, which form such breathtakingly beautiful carpets up in the mountains of Corsica or Greece have been bred into those large versions we give one another at Christmas. They, I admit, are lovely, but such cultivars can only too easily become blowsy gaudy show-offs.

Not all wild flowers, of course, are modest and small. Some are truly gigantic. The largest of all, so large that it's really not unfair to call it monstrous, is something called *Amorphophallus titanum* – the titan arum. It is a relative of the little lords-and-ladies that you can find in an English hedgerow. This consists of a vertical shroud wrapped around a central spike. If you unfurl the shroud, you will see several dozen florets, clustered around the bottom of the spike. In the autumn, the shroud rots and reveals that the flowers have now developed into bright red berries.

The titan arum is much like that, except that its shroud is not cream-coloured but greenish yellow on the outside and a rich dark crimson within. And it does not stay furled around the spike but opens out into a huge cone five feet across, like an immense upside-down umbrella with its point sticking in the ground. And from the middle of it rises a spike that can be ten feet tall.

The titan arum grows in the tropical forest of Sumatra. But it is not easy to find – at least its flowers aren't. Young *plants* are easier – except for the fact that you can easily mistake one for a sapling tree. They stand fifteen feet tall with a glossy green trunk – or to put it more accurately, a stem – that is as thick as an average telegraph pole. At the top the stem divides into three arms, each of which carries lines of leaflets. The leaflets, of course, like all leaves, produce starches, the plant's food reserves, which are stored in the swollen top section of the root just below the surface of the ground, like the corm of a crocus. Except that the titan arum's corm is the largest in the world. It can be four and a half feet across, so big that it would take two men to carry it. By the time it gets to that size, the plant will be about six or seven years old. And then the giant tree-like leaf that produced it, dies and falls.

In the humid heat of the forest, the fallen leaf rots rapidly and within a few weeks there will be nothing on the ground to show where the corm lies. That remains dormant for three or four months. And then – at no particular time of the year, and with extraordinary speed – it starts to flower. First, a spike appears above ground, growing rapidly – almost visibly. Within a few days it is as tall as a man. And then the umbrella-like shroud starts to unfurl revealing a grey fluted spike rising from the centre which continues to grow until it may be ten feet high.

A few years ago, I set out to try and film one of these extraordinary flowers. We settled ourselves in a small Sumatran village and offered a reasonably generous reward to anyone who could show us one of these blooms which, it's said, produces a smell of carrion, and is therefore known locally as *bunga bankai* – corpse flower.

After a couple of days, a man rushed in, panting. He had found one. We grabbed our cameras and off we went. It was raining heavily – this after all is a rain forest – and the going wasn't easy. But after a couple of hours of staggering through the drenching forest, our guide stopped, pointed to a slimy brown mass on the ground and claimed his reward. 'But it's dead', we said. Well, we hadn't specified that it had to be alive, he replied. And anyway it had been in flower *yesterday*. So we paid up. And then, on the following day we did find one, just at the beginning of its three-day flowering.

Sitting beside it, at this stage, was – oddly – somewhat unnerving. The flower is so out of scale with its surroundings, so vastly bigger than any leaf or flower around that it didn't seem to belong there. It was almost as though it was some alien that had arrived from outer space. It even seemed to be moving. There was no wind, down on the forest floor, yet I could have sworn that the great crimson umbrella was pulsating slightly. And its smell also came in waves. – rather mushroomy, I thought, and not really the notorious stomach-churning stench of dead flesh.

What pollinates this gigantic flower? No one could tell us. But as we watched, we saw tiny sweat bees fly over the lip of the umbrella and go down into its depths. We cut a small hole in the base of the umbrella – I confess it felt like sacrilege to do so – and peered in with our cameras. Like lords-and-ladies, there were two bands of florets encircling the base – the female ones, each with a stigma like a matchstick projecting from a globular base at the very bottom and the males, a mass of tightly packed yellow rods, above. The bees, of course, were seeking pollen, but to get it they had to crawl down into the very depths of the umbrella. There, they pollinated the female flowers by rubbing off pollen that they had collected some time earlier when visiting another titan arum flower.

The following day, the umbrella began to close and by the evening its top edge had wrapped itself around the spike. And there, inside this huge parcel, protected from the rain and interference by any large animal, the female flowers would over the next weeks swell into scarlet cherry-sized berries.

But why does the spike have to grow to these huge proportions – such record-breaking heights? Well, the plant is not common – as we well knew. Nor do the flowers appear at any particular season. So the chances of two plants starting their irregular flowering simultaneously, on exactly the same three days, is not high. So when one *does* open, the nearest source for the pollen it needs might well be a mile or so away. So the flower has to produce a pretty powerful scent and make sure that it travels a long distance. When human beings want fumes from their factories to blow away in the wind, they build tall chimneys. *Amorphophallus* generates perfume – if that is what you can properly call it – and disperses it far and wide through the forest by releasing it from the top of its record-breaking spike.

Technically speaking, the titan arum is not the largest flower in the world – because it is an assemblage of tiny ones, its florets. The largest single flower, in the strictest sense of the word, however, also grows in the same Sumatran forests. It was first described scientifically when in the early years of the nineteenth century Sir Stamford Raffles, having founded Singapore, went on to bring Sumatra, for a few short years, into the British Empire. Having taken up residence there, he commissioned scientists to survey the island. One of them found this immense flower and it was named after Raffles himself – *Rafflesia*.

It's a very odd plant indeed. Like the titan arum, its flowers appear out of nowhere. They draw their sustenance not from a huge tuber in which they have stored the food they have manufactured over the years. Instead, they steal it from another plant – a species of vine. *Rafflesia* is a parasite.

For most of the time, it exists invisibly as a tangle of filaments within the tissues of the vine's stem. But every now and then, in places where the vine stem trails across the ground or is even half buried in it, a bump appears on the vine's stem. This swells very quickly, day after day, until it is the size of a cabbage. Then it bursts open. Great leathery petals, reddish brown blotched with cream, spread open and lie flat on the forest floor. In the centre of the flower there is a strange plate like disc raised on a pillar and covered with spikes. There are a number of species of *Rafflesia* but the largest bloom produced by any of them is almost three feet across. It is pollinated not by bees like the titan arum but by bluebottles that normally feed on carrion.

The titan arum, understandably, grows vertically in order to disperse smell. But why does *Rafflesia* grow so big – horizontally? Flies can hardly require it to be three feet across before they notice it. And in any case, it is the flower's smell rather than its appearance that attracts them initially. Pondering on the problem, it occurred to me that the size of a plant's flowers, in most cases, must be governed by those flowers' effectiveness. There is no point in using up valuable resources to construct an immense flower, if increased size does not attract an increased number of pollinators.

But that sort of economics doesn't apply to *Rafflesia*, because it doesn't manufacture its own food. It takes it – steals it, you might say – from its host, the vine. So it can squander its unearned income on bigger and bigger flowers at no cost to itself. And I thought about the great country houses of Britain – immense baroque structures like Chatsworth, say, or Holkham Hall. Could it be that *Rafflesia* should be regarded as the aristocrat of the plant world?

Sir Thomas Stamford Raffles, in addition to his duties as a powerful colonial administrator in Britain's south-east Asian empire, took a deep and scholarly interest in the natural history of the territories he controlled. On his return to Britain, he was instrumental in founding the Zoological Society of London. His name, with some justice, was given to the enormous flower that was discovered during his explorations of Sumatra.

Above left. On the first afternoon of its flowering, lords-and-ladies heats up and gives off a faint smell of dung. This attracts small flies. The central purple spadix is slippery and flies landing on it fall down into the globular chamber beneath and are dusted with pollen from a band of male flowers around the spadix. They also crawl over the female flowers that grow beneath in another band. As they do so the flies fertilise them with pollen that they collected earlier from another flower.

Above right. By mid-summer, the shroud around the spadix, which is known as the spathe , has withered and fallen away, exposing the red berries. Their brilliant colour attracts the attention of birds which eat them and then void the seeds elsewhere with their droppings. The berries of the titan arum are a similar colour, but the size of marbles.

Right. The flowers of the titan arum, clustered around the base of the giant spadix. The tightly packed yellow flowers at the top are male and produce pollen. Those beneath, each with a long style tipped with a yellow stigma, are female. At the very bottom, just above the cut edge of the spathe, is one of the pollinators, a small sweat bee.

The first specimens of *Rafflesia* were sent back to Britain pickled in casks of brandy. From one of these, Franz Bauer, the accomplished Austrian botanical artist who was working in the Royal Botanic Gardens in Kew, produced this remarkably accurate drawing which was published to illustrate the first scientific description of the species.

The species discovered by Raffles' party was not in fact the first plant of its kind to be described. Twenty one years earlier, in 1797, Louis Auguste Deschamps, a French naturalist working in Java, had found a similar monstrous bloom. France and Britain were at war at this time and Deschamps' notes were never published. It was not until 1825 that a scientific description of his discovery appeared with this strange highly simplified illustration. By now the generic name *Rafflesia* was established and had priority. Deschamps' plant was given *patma* as a specific name, this being the one used by the local people in Java.

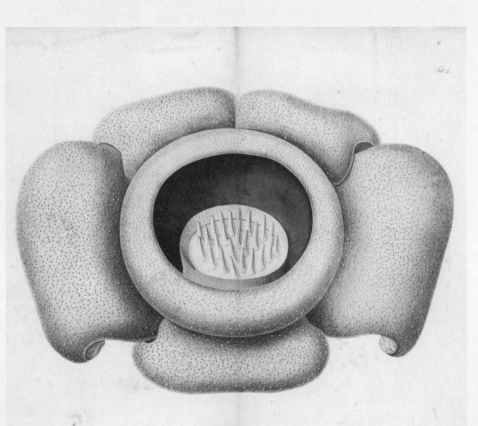

RAFFLESIA PATMA.

It may take a year for a bud to grow from its first emergence to its full size. It then begins to unfurl its of its huge leathery petals. Two days more, and it is fully open.

Rafflesia's berries are eaten by squirrels and, as here, by tree shrews. These animals doubtless transport the seeds, either internally or stuck to their snouts. But how do the seeds manage to get inside the stems of the vine in which they will germinate and grow? That is still a mystery.

Once fully unfurled, the flower of *Rafflesia* remains open for eight days. Then it begins to decay and in about two weeks it will have been reduced to black slime.

3

Platypus

Two or three hundred years
ago, there were quite a lot of
preserved mermaids around.
You could see them in
fairgrounds – if you paid your
money. They were about two
feet long, rather shrivelled and
very unattractive.

◁ A platypus swims with its eyes closed. The white mark behind its bill is not an
eyelid but a pale spot on its fur.

Two or three hundred years ago, there were quite a lot of preserved mermaids around. You could see them in fairgrounds – if you paid your money. They were about two feet long, rather shrivelled and very unattractive. Collectors occasionally bought them – for high prices. They were, of course, fakes, manufactured somewhere in the Far East by sewing the head and upper torso of a small monkey on to the back end of a cod or some such fish. Looking at them now – and you do sometimes come across them exiled to the store-rooms of museums – it is difficult to believe that anyone was taken in by them. But back then, people were rather more gullible than they are now.

So when in 1799 a dried body of a creature arrived from Australia that had the beak of the duck, the fur of a rabbit and four webbed feet, you can hardly blame a man of science for viewing it with some scepticism. It had been sent from the newly established colony in New South Wales, and Dr Shaw in the Natural History Department of the British Museum was understandably cautious. He examined it with great care. As far as he could see, the grotesque bill was attached organically to the furry skin of the head and not stitched on. The feet, with their webbing, also seemed to be perfectly genuine and not taken from some duck. So Dr Shaw accepted it, registered it in the archives and gave it a name. He called it 'platypus' – Greek for flat-footed, though I have to say that that seems to me the least of its anatomical oddities.

In due course, more preserved platypuses arrived in Europe. One went to Professor Blumenbach in Göttingen in Germany. He gave it another name because it had been discovered that Dr Shaw's name was not valid. Someone else had used 'platypus' as a name for a family of flat-footed beetles, and the rules of scientific nomenclature don't allow duplication. Blumenbach probably didn't know any of its aboriginal names – one of which I was charmed to discover is boondaburra – but coined one which was, you may think, rather more appropriate than Dr Shaw's. He called it *Ornithorhynchus* which means bird-billed and does therefore refer to the animal's most conspicuous and extraordinary feature. Scientifically speaking *Ornithorhynchus* is now its correct title but Dr Shaw's name has stuck. It remains platypus for most of us.

Anatomically, the platypus is an extraordinary mixture. Its most obvious anomaly, its beak, is not in fact very bird-like. It is soft and rubbery and covered in tiny sensory pores. Some of these it uses to feel its way around. Others, more remarkably are able to detect the minute electro-magnetic radiations that most living animals discharge. So when you watch a platypus underwater, swimming energetically along the river bed waving its beak from side to side, it ought to put you in mind, not of a duck grubbing around in the mud of the river bottom, but of a human treasure hunter, walking over an archaeological site waving his electronic metal detector.

The platypus's fur – a beautiful reddish brown – is soft and warm. Aboriginal people made cloaks from it and for a mercifully brief moment of fashion, Victorian ladies wore capes and muffs made of it. But European scientists, when they came to examine the animal in detail, found that its most remarkable feature was neither its webbed feet nor its duck bill, but something at the back end beneath its flat furry tail. The animal does not have two vents – one for bodily waste and the other for reproduction as a regular mammal has – but just one, like a bird or a reptile.

So the big question was, how did this extraordinary animal breed? Dissections revealed that the females had shelled eggs in their oviducts waiting to be laid. So to that extent the platypus was like a reptile. But then it was found that although these females did not have nipples, they did have enlarged glands in the skin of the underside, beneath the fur, that produced a fatty nutritious fluid – a sort of milk – and milk is a characteristic of mammals. So this extraordinary creature represented a link between the two groups.

Thirty years ago, I started work on a series of programmes in which we tried to trace the history of life on this planet using living animals to illustrate all the important stages of its evolution. We called the series *Life on Earth* – and of course one of the key events was that transition – from reptiles to mammals. Platypus provided the most vivid and dramatic way of illustrating that. But no one had ever filmed that key event, its breeding behaviour.

In the wild, platypus spend most of their time swimming in rivers and shelter in long burrows which they dig for themselves in the banks. It is in these burrows that the female lays her eggs. At that time it seemed that filming her – in her nest – in the wild – would be impossible. So what chance was there of persuading a platypus to breed in captivity and doing so in a nesting chamber into which our cameras could look? We put it about in zoological circles that we would finance a programme of scientific research into platypus breeding behaviour if we would be allowed to film it. But there were no takers. People pointed out, quite correctly, that platypus were exceedingly difficult to breed in captivity. It seemed as though I would have to describe this key characteristic in words while viewers watched film shots of platypus simply swimming about. And then, right at the last moment, when we were editing that particular programme, we discovered in an Australian Government film archive a shot of an echidna egg hatching. The echidna is the platypus's only close relation, a cousin that remained on land and protects itself with a coat of spines like a hedgehog. But it too lays eggs and suckles its newly hatched young on milk. It was a remarkable sight, the egg cracking and a tiny naked baby mammal, an echidna, peering out. True, it was hardly natural, because the egg was in someone's hand and the baby was already half way out when the shot started. But it was such an extraordinary sight that we put it in the film anyway with an appropriate explanation.

Years later at an official party in Australia someone introduced himself to me as the cameraman who had taken that echidna shot.

I thanked him for it and said what a pity it was that he had not been able to film the whole process. Did it happen unexpectedly and with no notice? 'Ar – no' he said. 'I filmed the whole thing from the beginning, but it was only when we were half way through hatching that I noticed I had not taken the lens cap off the bloody camera'. Such were the hazards of filming in the days before electronics replaced film.

But then, a few years ago, I had a chance to return to the problem. This time we *did* have electronic cameras and – more than that – we had endoscopes, fibre-optic cables with lenses at the end, such as are now routinely used by doctors to peer inside our bodies. And we also found an Australian researcher, Tanya Rankin, who was already studying platypus in a beautiful wooded valley in rural New South Wales. She had fitted a female platypus with a tiny radio transmitter that could send a signal sufficiently strong to pass through earth. With that she was able to trace the course of a female's burrow, which can be twenty feet long, and plot on the ground above exactly where the breeding chamber lay.

Platypus don't dig tunnels in the way that rabbits, for example, do. Instead of excavating the soil and then dumping it outside the tunnel entrance, they force themselves through the soft soil of a river bank, pushing it aside with their duck bill. As a result, the earth immediately around the tunnel is much more tightly packed than elsewhere. So when Mark, our Australian cameraman, produced an augur and started to bore a hole from the surface into the breeding chamber, he was able to tell when he was about to break through – and he didn't go the final half inch or so until he knew that the mother platypus was away feeding in the river. So the potentially alarming breakthrough happened when she was away. And when she came back we were able to watch her crawling into her chamber. And we saw more. In the tangle of plant stems with which she had lined the chamber, we saw a squirming white hairless creature almost as featureless as a slug. It was a newly hatched baby.

In a way that was a bit of a disappointment, for it meant that we had missed the actual hatching of the egg. But it was wonderful enough. Over the next few days, we filmed the mother platypus as she came and went. And eventually we saw the baby supping globules of milk that oozed from the female's furry underside.

After a week or so, we decided to stop filming – both for Mark's sake and the platypus's. Although we were fairly sure that she didn't notice our intrusion, we didn't want to risk anything. We would return in a few weeks' time when the baby was less like a white slug and more like a furry mammal.

But that never happened. A week or so later there were violent rainstorms all along that part of eastern Australia. The river in which we had watched the female platypus hunting for crayfish rose several metres and all the platypus burrows in the valley were flooded. When at last the waters fell and Mark and Tanya returned, the burrow was soaked and empty. So that magical moment, the sight of a platypus egg cracking open and a little mammal looking out from it, has still to be recorded.

George Bennett was both a medical doctor and an expert naturalist. He caught and kept platypus and in 1866 published this drawing of his excavation of one of their nesting burrows. He also dissected the animals and revealed that not only did the female have milk-producing glands on her underside but her oviduct sometimes contained a shelled egg. Whether this was laid before it hatched or whether it hatched internally so that the young were born unshelled, he did not know.

This mermaid was put on exhibition in Piccadilly in 1824. The accompanying booklet told customers that she was 'purchased by the proprietor in the archipelago of the Malacca at a very considerable price'. Her upper half was probably from a monkey and her lower from a fish. It was manufactured freaks like this that led scientists to question the authenticity of the first platypus specimens to arrive in Europe.

ORNITHORHYNCUS PARADOXUS.

AN AMPHIBIOUS ANIMAL of the MOLE KIND.

The first thorough and comprehensive account of the infant colony of New South Wales was written by David Collins, the Governor's Secretary and Judge Advocate, and published in 1798. In it, he describes some of its strange animals and includes a drawing of one of them made by the Governor himself, though he remains, very properly, somewhat cautious as to its identity.

In 1800, the transport ship *Minerva* arrived in Sydney harbour and stayed for a month. The ship's surgeon, John Price, went ashore and was lucky enough to see a platypus on land. The sight so impressed him that he drew this sketch of it in his journal. His European prejudices however led him to describe this astonishing animal as typical of the 'helpless, deformed and monstrous' creatures that are found in 'remote solitudes'.

John Gould, the great nineteenth century naturalist-publisher, in addition to the bird books that made him so famous also produced a volume on Australia's mammals, from which this illustration comes. Although his was one of the first really detailed first-hand descriptions of the platypus, even he did not believe that the animal laid eggs.

The platypus is truly amphibious for it regularly emerges from the water and nests in holes it excavates in river banks. It should however be handled with care for, uniquely among mammals, the male has a spur armed with venom on each of its hind legs.

The platypus' sole living relative, the echidna, is totally land-living and has developed a protective coat of spines like a hedgehog. And instead of a duck-bill it has a rod-like probe, to help it to collect ants from the ground. Internally, however, the two animals, superficially so different, are very similar and the eggs of the two species look almost identical. Here is the echidna's, with the hatchling's tiny foreleg, still only half-formed, reaching out of the shell.

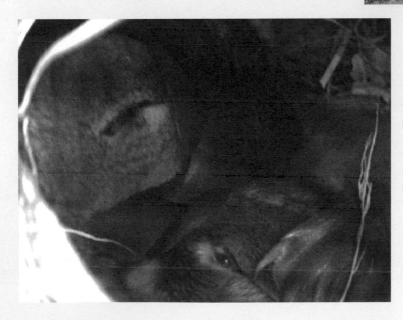

The Australian film-maker, David Parer, using an endoscope, took this photograph of a female platypus curled up around her baby. The youngster was sufficiently old to have developed its duck-bill but was still feeding on the milk that oozed from its mother's fur.

4

Giant Birds

The biggest egg ever laid – by anything – is about the size of a rugby football. It is so big that it can hold about a gallon of liquid – say a hundred chicken eggs. The first reports of such monstrous eggs were brought back to Europe by Marco Polo.

◁ Most of the eggs of *Aepyornis*, of course, hatched. Only those that were addled or deserted remained unbroken. So when local Malagasy people find one they have discovered a rare and valuable treasure.

The biggest egg ever laid – by anything – is about the size of a rugby football. It is so big that it can hold about a gallon of liquid – say a hundred chicken eggs. The first reports of such monstrous eggs were brought back to Europe by Marco Polo, the Venetian traveller who reached the court of Kublai Khan in China at the end of the thirteenth century.

There, he said, he had seen one of these astounding eggs, together with a couple of the feathers of the bird that had laid it. The feathers were probably the withered leaves of palm trees, but the egg, it seems, was real enough. He did not claim to have seen the bird that laid it, but he was told that it was so huge and powerful that it could kill elephants by picking them up with its talons, flying up to a great height and then dropping them. The name of this alarming creature was the rukh or roc and it lived, he was told, on an island south of Mogadishu, a city on the east coast of Africa where Chinese merchants went to get elephants' teeth – that is to say, ivory

Well, south of Mogadishu there *is* a great island – Madagascar (the names are so similar there may in fact be some confusion between them). It is a thousand miles long and in the seventeenth century a French explorer, Flacourt, reported that there *was* a giant bird living in the south of the island that laid enormous eggs. But it was not an eagle, he said, it was a kind of ostrich.

By the nineteenth century, Madagascar had been sufficiently well-explored to establish that this large bird was extinct. But the eggs were still around and three of them were brought back to Paris by a French traveller. Marco Polo could have been right about the eggs, at least. They are still the biggest eggs known – much bigger than those laid by dinosaurs, the biggest of which are only the size of a grapefruit. And they are probably as big as an egg can get, because the bigger the volume, the thicker the shell has to be to hold its contents – and there is a limit to what the poor chick inside, armed only with a little spike on the end of its beak, can be expected to hammer open in order to get out.

Some forty years ago, I was down in southern Madagascar, filming lemurs. One evening, after we had pitched camp, I went for a stroll along a dried-up river bed and noticed what looked like a small fragment of ceramic tile – about a quarter of an inch thick, slightly curved, rough on one side but pitted and a pale yellow on the other. I remembered those elephant bird stories and was soon down on my hands and knees, scrabbling in sand. And after some searching, I found three or four more bits. All very exciting!

The following morning, a lad driving a herd of goats wandered by the camp. I showed him the fragments of what I hoped were elephant bird eggs and said I would give a good reward for any more that he could find. He looked at me with a baffled expression and drifted away. I doubted that he had understood a word I had said. But that evening, a woman came into camp carrying a basket on her head. She tipped it out in front of me and out poured what I suppose must have been about five hundred or so pieces of egg shell. Fortunately, I hadn't managed, in my inadequate French mixed with a few Malagasy words, to make myself sufficiently understood to the little boy, to commit myself to a price per piece. If I had, I would have been bankrupt, because after that first one, a procession of women arrived at our camp, each bringing basketfuls of egg shell fragments. By nightfall I had a huge pile of them a foot or so high. Which just goes to show how incompetent I was at spotting fossils. It might also suggest how abundant the elephant birds must have been, except that, when you come to think about it, whereas chicken eggs or those of any normal bird are so thin they crumble to nothing in no time at all, these elephant bird eggs had shells that were so thick that after they had hatched, the fragments didn't disintegrate but simply lay around in the sand – in millions.

And then, just as I was wondering what on earth I was going to do with such a multitude of fragments, the little goatherd reappeared. He too had some pieces but these were really large, some nine inches or so long. As he laid them out on the ground, I could see immediately that a number of them fitted together. I got to work joining them with sticky tape and in an hour or so I had reassembled a virtually complete egg. And it was indeed as big as a rugby football.

Skeletons of the giant bird itself have now been found, so we know that *Aepyornis*, to give it its proper scientific name, was indeed ostrich-like. And that is hardly surprising. A bird big enough to lay such a whopper of an egg must have been so big and heavy it would have found it very hard indeed – if not impossible – to get airborne.

But why should any bird abandon flight? Well, flying is a very energetic business – much more so than walking or running – and birds don't fly unless they have to. Indeed if it is safe to stop, that is what they do. The process has occurred repeatedly in recent times on islands where there are no ground predators. So cormorants on the Galapagos and rails on several Pacific islands have stunted wings that are no longer big enough to get the bird into the air. The same thing happened, world-wide, around sixty five million years ago when the dinosaurs – for some reason or another – vanished. At that moment in the earth's history, it was relatively safe on the ground and some birds abandoned flight. And they did more. They made a bid to dominate the land. The only mammals around at the time were small furry rat-like creatures. The flightless birds, however, with no need to worry about their weight, grew very large indeed. Some became really fearsome hunters with huge beaks like meat-cleavers with which they slaughtered smaller creatures such as lizards and – presumably – the early little mammals.

But the mammals didn't remain subordinate for long. Eventually *they* won the competition for dominance. A few of those flightless birds that made a bid for the throne still survive – the ostrich in Africa, now the biggest of all living birds, the rheas in South America which are much like them though somewhat smaller, and the emus and cassowaries in Australia.

Aepyornis was one of them that evolved on the continental fragment that is now Madagascar. It probably survived until the seventeenth century, for Flacourt the Frenchman wrote as though it were still alive. New Zealand, too, had its quota. Indeed, it had rather more than any other continental fragment, for it broke away from the supercontinent very early, even before the mammals had started their rise to dominance, and none of them ever reached it. So New Zealand had not just one species but a dozen or so of these ancient flightless birds. They were known as moas and there was a range of about a dozen different species of them. They survived in the safe isolation until around a thousand years ago when human beings, Polynesians, arrived there. These new immigrants cleared much of the forests and hunted the moas relentlessly – and by the seventeenth century, all the moas were extinct.

In general shape, the moas were not so very different from ostriches. And some were very big indeed. In fact, one of them might well have stood even taller than Madagascar's elephant bird. But there is a problem about determining the heights of moas. In the nineteenth century, when their bones were first being unearthed, scientists and collectors competed to collect the tallest species. A major part of the moas' height, of course, came from their neck and if you happened to excavate a more or less complete skeleton and then found a few odd neck bones nearby, why not add them to the neck of your skeleton and give it a few extra inches? And it is now pretty clear that some skeletons of what were once thought to be record-breakers contain neck bones from more than one individual.

And then there is the question of posture. Did the moas hold their heads upright like an ostrich or rather lower, with the neck in a curve, like a cassowary? We are not sure. But it does seem that, even if the elephant bird was not taller than any moa, it was certainly heavier.

And on one point there is no debate at all. Moa eggs are really quite small – little more than half the volume and weight of the elephant bird's. So Marco Polo was quite right to be astounded by the immense eggs he was shown in the court of Kublai Khan. They are wonders. There *are* none bigger.

A goose's egg, a hen's egg and the egg of *Aepyornis*.

Cormorants on Fernandina Island in the Galapagos have no need to fly, for they are not threatened by land predators. As a consequence their wing-feathers have become reduced and somewhat tattered, and cannot support them in the air. Their tail-feathers, however, remain broad and strong, for the birds use their tails as rudders when swimming underwater in search of fish.

Ferdinand Magellan, the great navigator whose expedition was the first to circle the globe, is seen in this sixteenth century engraving sailing through the Straits at the southern tip of South America that now carry his name. He is busy with his navigational instruments, a compass, dividers and an armillary sphere. The broken mast on the deck indicates the storms that battered his ships. Apollo, the god of music, is hovering alongside singing the explorer's praises.

On the nearby shores, there are the fires made by the inhabitants that gave the land the name it still carries of Tierra del Fuego, the Land of Fires. Monsters swim in the sea and a mermaid rises from the waves grasping her tail. On the cliffs to the right sits one of the giants that the expedition reported, thrusting an arrow into his throat so that he may stave off hunger by eating his meal a second time.

Above, on the right, the God of the Winds sits on a cloud and on the left flies the gigantic rukh bird, carrying an elephant in its talons.

Richard Owen was the most brilliant of Victorian comparative anatomists. It was he who first showed that giant reptiles – for which he invented the name 'dinosaur' – had once roamed the earth. He also deduced that a fragment of thighbone sent to him from New Zealand came from a giant extinct flightless bird. Here, photographed in 1877, he stands with the thigh bone fragment in his hand, beside one of the first moa skeletons to come to London's Natural History Museum of which he was the founding Director.

The ostrich, like *Aepyornis*, is flightless. Its eggs are the largest laid by any living bird but, in terms of volume, they are only about a sixth of the size of an *Aepyornis* egg. The male ostrich establishes a nest site and then as many as half a dozen females lay eggs in it. He may thus have twenty or so eggs in his charge. Whether the elephant bird had a similar habit we do not know.

The eggshells of most birds, once the young have hatched, quickly crumble to dust. *Aepyornis* eggshell, however, is so thick that it does not do so and in consequence vast numbers of fragments can be easily collected from the desert sands of southern Madagascar – if you know where to look.

5

songsters

The human voice is very odd.
I'm not thinking of its ability to
produce yelps of pain or
threatening roars, or even to
communicate with speech. I'm
talking about singing.

◁ The siamang lives in the Malaysian Peninsula and Sumatra. Like all gibbons it
is an accomplished singer. Both males and females have large throat pouches
which amplify their calls to such an extent that they can be heard for miles through
the forest. Male and female sing together both in the early morning and at
sundown. Their dramatic choruses seem to silence all other creatures in the
surrounding forest.

The human voice is very odd. I'm not thinking of its ability to produce yelps of pain or threatening roars, or even to communicate with speech. I'm talking about singing – a tenor producing an eye-popping and hair-raising high C; or a soprano executing perfectly articulated runs and trills. Or, come to that, several thousand well-lubricated throats in the National Stadium in Cardiff singing together and in harmony, *Bread of Heaven*.

Evolutionary theory maintains that physical characters develop because they are needed, because they bring their owners an advantage. But what need was there for the human throat to have evolved a larynx that can produce such pure, musically accurate sounds over such an astonishing range? Whatever that need, it must have been a very long time ago to allow such a complex anatomical device to develop. It may even have existed among our far-distant humanoid ancestors, even before our species, *Homo sapiens*, appeared on earth.

Human beings, of course, are not the only creatures that sing. Birds do. And again, I'm not thinking about the chattering notes of an outraged blackbird or the harsh aggressive squawks of a magpie. Such vocalisations carry simple messages – 'look out, there's danger'; or 'this is who I am and this plot of land is mine.' No. I am talking about real *song* – sustained, complex and melodic. The songs of birds have characteristic scales, durations and intervals that makes it possible to identify the species of the singer. They also may differ sufficiently in their details for even a human ear to recognise a particular individual. It is likely to be a male, for song is an essential part of his courtship. Bird song is connected with sex.

Interestingly, there seems to be a rough relationship between song and plumage. The more visually stunning a bird is, the simpler and – often – the harsher its call is likely to be. So when you see a parakeet, resplendent in green, blue or dazzling scarlet, you can predict before it even opens its beak, that it will make a harsh sound and one so simple it could hardly be described as a song. A male mandarin duck, the most extravagantly bedecked of all ducks, with epaulettes, ruffs, and mantles, can only produce what even a duck-lover calls, technically, a 'grunt-whistle'.

The most extreme example, perhaps, is the peacock. Few birds can outdo the shimmering splendour of his erected tail feathers, but that visual revelation is accompanied by a shriek that was once described to me as the equivalent of being jabbed in the ear with a screwdriver.

Conversely, the most accomplished singers are very often the plainest of birds. Which of our British birds has the most complex and beautiful of songs? Many would say the nightingale. And what does it look like? Well, of course, it is plain brown with pale under-parts – a costume that could hardly be more modest.

Nature seldom goes in for both belt and braces. So you might well guess that the display of bright feathers and the carolling of a complex song are alternatives – different ways of achieving the same goal. Both have a sexual purpose. Both are ways by which a male bird attempts to persuade a female to mate with him rather than with a rival.

There is a famous if a somewhat unfeeling experiment that was carried out on the African widowbird. It lives in the grasslands of Kenya. Out of the breeding season, both sexes look like rather large sparrows. But when the breeding season arrives, the male grows black feathers and a long glossy black tail, and sets up his territory amongst the tall grass. The experimenters caught a range of these males. They shortened the tails of one group so that they were only about six inches long. Then they added those cut pieces to the tails of another group so that their tails became about twice the normal length. And yes, it really did work as you might have predicted. The trimmed males postured with just as much self-confidence and bravado as before – but found themselves largely ignored, while on the other hand, those with lengthened tails found themselves overwhelmed by female admirers.

Well, song is the audio equivalent of the widowbird's tail. The longer and more complex it is, the more successful it is. The most dramatic proof of that, that I know, has been produced by a group of Swedish ornithologists working with great reed warblers.

These birds are summer visitors to Sweden. First to arrive in spring are the males. Each establishes a territory for himself in the reed beds by singing duels. The birds are so pumped up with testosterone that these duels can go on, one after the other, all day, which here in the north at this time can last for twenty hours. The songs, at this stage, are relatively simple. Each contains a number of different syllables – between ten and fifteen – which are produced in varying combinations. By duelling in this way, the males space themselves out in more or less evenly spaced territories across the uniform expanse of reeds. Then, a week or so later, the females arrive from Africa and immediately the males change their calls from signals into songs – longer, more sustained and more varied.

The females tour the territories, listening to different males. To our ears, the songs of the different males may sound very similar. But if you record them and slow them down, you can count the different syllables that any one song contains. And they vary quite considerably. Good performers can produce songs that contain 45 syllables; bad singers only 25. And a female, when given the choice, invariably mates with the one with the more syllables.

And what about mammals? Do mammals – apart from human beings – have songs? Well, there is one group which really does produce songs that in melodic invention, excitement and complexity, outshine all rivals – except mankind. You are likely to hear it before dawn, ringing out over the canopy of the forests of the Far East – and once heard, never forgotten.

It is of course the song of the gibbon. There are several different species of gibbon – and they all sing. A young male, having left his family group to start life on his own, usually gives a solo recital in the early mornings. It can last for as long as a quarter of an hour and can be heard over half a mile away. Young females, on the other hand, don't sing at all. But they are clearly attracted by the male's song. I don't know of any studies that have been done analysing the different skills of individual young males, but I wouldn't mind betting that they *do* vary and that the female gibbons – like the female reed warblers – are discerning critics.

It may well be true that the fitter, the stronger and healthier a male gibbon is, the better he sings. But I don't believe that the females say to themselves – that is a wonderful, powerful, inventive song and therefore, if I mate with the singer, my offspring will be stronger and healthier than if I mated with a less talented singer. No. She is simply beguiled by his singing. She has, in fact, a very real sense of musical appreciation. So when she hears a really accomplished singer, she is delighted and she joins him. Singer and listener don't always click immediately. The male may have one or two liaisons but eventually the two form a pair bond. And now – she sings with him. They duet!

They practise singing together, working out the details of their joint performance until they have created an elaborate composition that is personal to them. It starts with a warm-up in which the male and the female exchange short phrases. But then, usually, the female takes over and embarks on a magnificent solo aria – her great call – to which the male may add encouraging interjections until finally the two of them join together in a wildly passionate climax, whooping and calling as they fling themselves through the branches high in the canopy of one of their favourite trees. The whole performance may last for a quarter of an hour.

In due course, after an eight-month pregnancy, the female gives birth to a single baby. It will remain dependent on its parents for some considerable time and be still with them when a second or even a third child arrives. As their parents continue with their music making, so the older among the young may join in. Singing, it seems, not only creates a family in the first place but is instrumental in keeping it together, even when the young arrive.

It is tempting to think that human beings, very early in their history, used music in a similar way. And there can be little question that a male with a good singing voice, in our own society, is still a source of sexual attraction and excitement. What else is a serenade? Watch a pop concert. And, just as among the great reed warblers, quality counts. Females have been selecting males with a versatile larynx since way back in our ancestry.

Today young men sing together to generate camaraderie; and religious people use song to generate the deepest and most profound emotions among themselves and their listeners. But the prime function of song is something else. Shakespeare wondered if music was the food of love. Well, vocally at least, it certainly was. And what is more, it still is.

The nightingale has always been much loved by Europeans for the ravishing complexity of its song with which the male invites a female to join him. This illustration comes from the first systematic catalogue of British birds published in 1676 by Francis Willughby.

The Great Reed Warbler is significantly bigger than the Common Reed Warbler but does not breed in Britain. The song of the male is remarkably loud and carries for around a kilometre. It is also very complex and it is this quality that persuades a female to chose one male rather than another.

The male Long-tailed Widowbird only develops his dramatic black plumage and long tail during the breeding season. At other times he, like the female of the species, is not unlike a brown sparrow. But she assesses his suitability as a mate by the length of his display tail.

The peacock is not an accomplished vocalist. It shrieks. The male has no need of vocal splendour. He attracts his mate by displaying his spectacular tail. The more 'eyes' in his train, the more successful he is likely to be in attracting females. The females are certainly very selective. Here one seems relatively unimpressed by this particular male's splendour and is moving away to see what else is on offer.

Although the human male at various times in history has dressed himself in flamboyant ways, he is not, anatomically-speaking, particularly decorative. But he does have a very versatile and powerful larynx which can be used to great effect as a sexual attractant.

6

Bowerbirds

There is an Australian story
about a one-eyed man,
wandering through the desert
who, before he stretches out on
his swag and goes to sleep,
takes out his glass eye and puts
it on a nearby tree stump. But
in the morning, when he
wakes, his eye has gone.

◁ The Great Bowerbird is found all over the northern third of Australia. The male
favours light coloured objects as decorations for his bower. Usually these include
snail shells and bleached bones but as long as they are white their precise nature
seems of little relevance. So males building their bowers close by a municipal
cemetery take marble chips from graves and others collect white plastic bottle tops
from refuse dumps. They also collect, in rather smaller numbers, green or reddish
objects such as fruits and leaves. The bower itself, in front of which these objects are
placed, may contain as many as four to five thousand sticks, each of which has had
to be carefully positioned.

There is an Australian story about a one-eyed man wandering through the desert who, before he stretches out on his swag and goes to sleep, takes out his glass eye and puts it on a nearby tree stump. But in the morning, when he wakes, his eye has gone. Who can have taken it? The bush-whacker, being wise to the ways of the wild, knows that there can only be one explanation. He walks off into the bush nearby, looks around and eventually finds an extraordinary construction beside a nearby rock – two parallel walls of twigs, a couple of feet long and about a foot apart. At one end there is a pile of white objects – quartz pebbles, bits of bleached bone, and some snail shells. And there, in the middle of them, staring balefully up at him – his glass eye. The thief – if that is what you want to call him – was a male bowerbird. And his walls of twigs were not a nest. They were a kind of structure that is unique in the animal kingdom – a sort of show-case, a bower.

A male bowerbird spends most of his time working on it – either building the walls or collecting fresh treasures to put in front of it. And all to impress the females. They, at the beginning of the breeding season, tour all the bowers in the district. Every time a female appears nearby, the male goes into paroxysms of strutting and shrieking, calling attention to his collection. But the females are very choosy. If one is sufficiently impressed, she moves into the bower and the two may mate there or, in some species, flutter off together into the nearby bush. Then away she goes, back to the nest that she has built *for* herself and entirely *by* herself. There, unaided by him in any way, she will rear her chicks. He meanwhile continues doing what he can to add to his treasures and the splendours of his bower – and hopes that his luck will continue.

He is quite pernickety. The bird that took the glass eye was probably a Great Bowerbird because that species prefers white objects. Other bowerbird species have other tastes. The Satin Bowerbird, which lives in somewhat better vegetated parts of eastern Australia, has a passion for blue. It is probably not a coincidence that he himself has glossy blue plumage and a glittering blue eye. It is as though the blue things he collects for his bower – blue berries, even bits of blue plastic, if he lives near human habitation – become extensions of his own body. Certainly, they seem to have the same effect as the extravagant plumes and decorations that are developed each breeding season by birds such as pheasants, ducks and birds of paradise. In fact his objects are rather more convenient than feathers because he doesn't have to grow them afresh every year. Nor does he have to carry them around with him all the time.

On the other hand, no male bowerbird is happy to leave his bower for long. He has competitors. If he is away, there is always the danger that one of his neighbours will nip in and steal some of his painstakingly assembled treasures. Worse, one might come and actually vandalise his bower, tearing down the twig walls with his beak – so as to reduce the competition.

Avenue-shaped bowers are not the only kind built by these amazing birds. Away to the north, in the great island of New Guinea, there are even more species than there are in Australia – a dozen of them – and the bowers *they* build are perhaps even more extraordinary. The first Europeans to see the constructions of McGregor's Bowerbird in the thick humid tangled rainforest could only imagine that they were the work of local people – something to do, perhaps, with their bizarre rituals. Each consists of a sapling around which is constructed a tower of interwoven twigs that may be four or five feet high. Around the base, even more remarkably, there is a kind of circular runway, the outer wall of which has a smooth rounded top made of specially planted moss decorated with little bits of evenly-spaced black fungus like raisins. And there are treasures here too. They are not quite as obviously attractive as the white pebbles or blue plastic of the Australian species but something more arcane, that can only appeal to a really sophisticated taste. Frass – caterpillar droppings. The male collects them from the forest and attaches them to the ends of the twigs of his tower.

When a female comes by, he hides. If she alights on the wall of the runway, he will scuttle around to the other side, so that he is hidden behind the tower, and start to sing. She cocks her head, listening, and he peeps out to see how she is reacting. If she thinks well of it, she will hop down into the runway. That is a very good sign indeed and he whizzes round and executes an excited strutting dance before the pair fly off into the bushes.

But even this maypole-like tower is not the most elaborate of bowerbird constructions. To find that you have to go to the western end of New Guinea known, because of its shape on the map, as the Vogelkopf, the Bird's Head. The bower built by a Vogelkopf Bowerbird is, to my mind, one of the most extraordinary sights in all nature. Its core is a tower of twigs built around a sapling like a maypole bower. But this is elaborated with a thatched roof stretching from the top of the maypole to the ground so that it looks like a low wigwam, with an entrance on one side that can be a couple of feet across and a foot high. The bower is so big in fact you can almost crawl into it. And that is not all. In front of the entrance, there is a meticulously tended lawn of green moss on which the bird has carefully arranged his treasures – glowing red leaves, acorns from a tropical oak, tree fern stems, or iridescent beetle wing-covers – each in its own separate pile.

Not all the Vogelkopf Bowers have the same kind of decoration. Some males clearly prefer particular colours. A bush that has just come into flower may provide a male with a speciality all his own. Another bird may try to impress with a collection of a particular fungus – and sometimes one so rare that he has to bring it from over a mile away.

Watch the bird when he arrives with his latest trophy and you will see that he doesn't just dump it on its appropriate pile. Not at all! He puts it down, hops back and looks at it critically, cocking his head, and then, maybe, hops forward again, picks it up and puts it back in a slightly different position. You have to think that he is considering whether or not it is correctly placed – from the point of view of its overall decorative effect.

Each male, it seems, is trying to impress a female with his own special arrangement. And the male that pleases her most will be the one she chooses to be the father of her chicks. She reaches that decision by making an aesthetic judgement – by choosing the display which pleases her most.

To film at a bower, you will have to build a little hide a tactful distance away. You must get into it before dawn so that the bird does not see you arrive. And then you must wait. A male will almost certainly appear and start to rearrange his treasures. Females, making their tours of inspection, are likely to make only a cursory visit and then leave. But if your film sequence is to be complete, you will need to end it with that last climactic shot in which a female decides that this is the male for her. She will make that clear by flying down to have a close look at the treasures on his lawn. Then the male, visibly thrilled to ecstasy, rushes out of the wigwam, and the two mate surrounded by his jewels. Filming *that* is a different matter altogether. To succeed, you yourself will have to tour the bowers and make up your mind as to which is the most attractive. It's rather like going to a flower show and trying to decide which entry is going to win the first prize. If you get it wrong, then you may in fact sit in your hide for ever and still not get that last shot. But if you get it right – if you pick the one bower in perhaps a dozen in the forest around you that was made and decorated by the champion male – then you will get that last crucial shot. And the fact is, that the chances of you doing so are quite high. It seems that we and bowerbirds have the same aesthetic sense and preferences. And that thought pleases me no end.

This is the earliest-known depiction of a bowerbird's bower. It was drawn in 1861 by one of two sisters, the Misses Harriet and Helena Scott, who were both enthusiastic amateur artists.

A male Great Bowerbird adds yet another snail to his collection. He scours the land around his bower for such treasures but he dare not go far afield, for if he is away for a long time, rivals in the neighbourhood may notice his absence and take the opportunity to steal some of his treasures.

Although a male Great Bowerbird lavishes so much time on his bower display, he still has some physical adornment, a little violet-blue crest at the back of his head. Normally this is kept closed and is scarcely visible, but if a female shows signs of interest and approaches the bower, he becomes very excited and encourages her by erecting it.

Constructions like this one, standing in the thick forest of New Guinea far from human habitation, baffled the first Europeans who saw them. It is clearly meticulously tended. The circular runway around the central maypole is immaculate and the surrounding wall is decorated with carefully-placed black fungi. It was made by a male MacGregor's Bowerbird as a stage on which to perform his displays.

Theft is not the only disaster that can affect a neglected bower. A rival male, in this case a Satin Bowerbird, may not merely steal the sticks from which the bower is made but deliberately damage it, pulling out tufts of stick, wrecking the walls, and leaving it in such a condition that, presumably, a female would not be attracted by it.

The Vogelkopf Bowerbird builds the most elaborate of bowers and uses the greatest variety of decorations. Each individual male seems to have his own particular visual taste and readily takes advantage of what decorative materials become available in the forest. This male is collecting acorns from a tropical oak.

The extraordinary bower of the Vogelkopf Bird. In construction it seems to be an elaboration of the maypole type, the top of which has been extended outwards so that it reaches the ground. It is not only the colour and shape of the treasures the male bird assembles that is important. It is their placing. Each kind has its own pile and a male will take great care in the placing of each item, putting it down and picking it up to try a new position several times before he is satisfied.

7

Dragons

The first really comprehensive encyclopaedia of natural history, published in Italy at the very end of the sixteenth century, has an excellent detailed section on dragons.

◁ The largest living lizard, the Komodo Dragon, is found not only on Komodo in eastern Indonesia but also on the neighbouring islet of Rintja and the western end of the much bigger island of Flores. The largest of them grow to around ten feet.

The first really comprehensive encyclopaedia of natural history, published in Italy at the very end of the sixteenth century, has an excellent detailed section on dragons. They are properly classified. There are six-legged, four-legged, and two-legged but with a pair of wings.Then there are those that breathe fire, and those that don't. The author, Ulysses Aldrovandus, was clearly in no doubt about their reality. What were they?

Well, clearly they were not crocodiles. Nor were they water monitors, those huge lizards that can grow to six feet long and can be very formidable indeed. Both seem pretty good candidates. But they weren't dragons, because people in the sixteenth century knew all about these creatures and Aldrovandus includes relatively accurate sections about them in his book. So I suppose that we must write off his reports of dragons as perfectly rational attempts to turn travellers' exaggerations into terms of flesh and blood. And indeed, dragons disappeared from natural history books within a few decades of Aldrovandus writing about them

But then three centuries later, in 1910, they reappeared. Reports came from what was then the Dutch East Indies and is now Indonesia, of a population of gigantic lizards, far bigger than any other species known, living on a tiny island called Komodo in the far east of the archipelago. And the popular press decided almost immediately that they, after all, must be the long lost dragons.

In truth, the Komodo Dragon is only a slightly larger version of the water monitor to which it is closely related. Initially there were claims that it was immense – twenty feet or so long – but the biggest actual specimens that anyone managed to collect, alive or dead, were only half that length. Then in 1926 a group of adventurous Americans sailing round the world visited Komodo and brought back photographs of the animal in the wild. By the 1930's one or two zoos had specimens. And then the story rather faded.

In the 1950's I was a young television producer, looking for suitable subjects. As far as I could discover, there was no film of the Komodo dragon apart from the rather fragmentary record made by the American expedition. Why shouldn't I try to film it? But when Charles, my cameraman companion, and I got to Indonesia no one in the main island, Java, seemed to have heard of the animal. Eventually, however, after some uncomfortable encounters with whirlpools, smugglers, and getting nearly shipwrecked on a coral reef, we got to Komodo.

To be honest, the dragons seemed to me to be quite docile. They were of course pretty impressive. Ten feet long may not sound much, but as one prowls towards you, the sun glinting on its grey shiny scales, and its huge yellow forked tongue sliding in and out of its jaws, savouring the air, it seems quite big enough. Even so I thought I would be able to fend it off with a pole if it got too inquisitive.

At that time, no zoo in Europe or America had one of these creatures in captivity. So when we came back, our film created quite a stir. Since then many camera teams and groups of scientists have gone to Komodo and made long-term studies to try and answer some of the questions that these remarkable creatures provoke.

First of all, are the dragons a relict population of giant lizards that were once much more widely spread in this part of the world – monstrous hang-overs, as it were, from the age of the dinosaurs? Well, there were once immense lizards living in Australia. Their fossils are quite well known and some of them were truly gigantic – up to twenty feet long and weighing over half a ton. They survived until around 25,000 years ago. But it is thought by scientists who have studied the question that Komodo's dragons almost certainly evolved separately on the islands where they now live.

Second, what do they live on? The answer to that seems to be quite easy. If you want to attract them, as we did, all you need is the carcass of a goat, and the more strongly it smells the better. But are the dragons only carrion feeders or do they, on occasion, become active hunters? And are they dangerous?

I asked in the island's only village if the dragons would attack people. Well, they said, one of the villagers some years earlier had disappeared and his body was found in the bush badly mauled by dragons. But they said, he *was* rather old. Maybe he had collapsed and died from natural causes and the dragons treated his dead body as they would the body of a goat. But then other people claimed that the dragons *did* actively hunt and *were* dangerous. They knock a goat – or a man – off his feet with powerful sweeps of the tail and then set upon their victim. I never thought that this was very plausible. But now it has been discovered that they are indeed killers and that they kill their prey in a rather different way.

Like all reptiles they take their time. They can sit around waiting for hours and days. Eventually they get a chance. A buffalo wanders close enough for them to lunge and deliver a powerful bite on one of its legs and immediately retreat. A kick from a buffalo could be lethal. Severe though the bite is, it is seldom enough to cripple their victim. But their saliva contains a poison. Slowly, it spreads through the buffalo's body, and slowly the animal gets weaker. And the dragon waits. And before long it manages to get another bite. And then another. After a week or more, the buffalo collapses. And the dragons move in and finish it off.

So they *are* hunters and they *do* live on buffalo and goats, alive or dead, as well as on deer and pigs. But there is a problem with this answer. None of these prey animals are native to the islands here. All have been introduced by human beings in relatively recent times. What did the dragons eat before then? This is less easily answered. But it seems that the dragons of Komodo are not just hunters – they are cannibals.

Newly-hatched dragons, only about eighteen inches long, are swift and agile and they feed on other smaller species of lizards which abound on the island as well as rodents, birds and even insects. But food is short on the island. So they are in danger of being eaten by the adults. As long as they are small, they can escape by climbing into trees where the large heavy adults can't follow them. But eventually they grow so big that they have to take their chances on the ground. There, the smaller ones will be eaten by the adults. Only those young that grow large – and quickly – will survive. That may well be the reason why this particular lizard, over many generations, has grown so big. It is a heartless technique that enables big animals to live on insects and other small items of food at second hand – by eating their offspring that are small and agile enough to catch and survive on those smaller things.

And there is one last question about today's dragons – and one last surprising answer. In recent years a number of zoos have acquired specimens and have done their best to persuade the dragons to breed. But it has not been easy. Indeed there have been tragedies. In London, a male dragon was introduced, with all the proper precautions, to a female who was already well established there. But she didn't take to her mate. Eventually, there was a scuffle. The male was badly injured and although it was rescued, it died. Other zoos had similar difficulties and tragedies.

The dragons had kept one of their secrets until the very last. Females don't need males in order to reproduce. They can lay fertile eggs without any contribution whatever from a male. They are, to use a technical term, parthenogenetic. And what's more, judging from the Zoo's experience, the females seem to prefer things that way.

Parthenogenesis does in fact occur in other species of lizards. One population of Mexican whiptail lizards, for example, *only* reproduces in this way. Initially, of course, it is a way for a species to take over a territory very quickly because every member of the population can lay eggs – instead of only half of them. But that success is unlikely to be permanent in the long run. A parthenogenetic female is only able to produce identical young. Only if a male contributes to the genetic make-up of the young will there be variation. And without variation there can be none of the adaptation that is necessary if a species is to survive environmental change. And if it can't do that, then the last of the dragons will indeed disappear.

Belief in the existence of dragons was still widespread in the fifteenth century. Whether Leonardo da Vinci, with his profound curiosity about the natural world believed in them we do not know, but certainly, there was a professional requirement for artists to paint dragons being attacked by saints. His detailed understanding of animal anatomy enabled him to draw one that looks remarkably convincing.

The Italian encyclopaedist Ulysses Aldrovandus devotes fifty pages in the second volume of his great thirteen part encyclopaedia to dragons of various kinds. This one, with bull's horns, a snake's tongue, scaly wings and a snake's tail comes, he says, from the wilds of Ethiopia.

Young Komodo dragons are markedly different from their parents. Instead of the uniform grey, they are quite brightly marked with yellow stripes. They are also very nimble and if chased by an adult they can escape by swiftly climbing trees where the heavy cumbersome adults cannot follow.

Komodo dragons could not bring down such a massive animal as a water buffalo in a straight fight. Their hunting technique is less direct and more long-drawn out.

They lurk behind a buffalo and when they get the chance they snap at its hind legs. The bite in itself is not severe. But it carries a poison. With reptilian patience, they trail their victim. As it weakens so they get the chance for further bites and eventually, after maybe a week, the animal collapses, and the dragons feed.

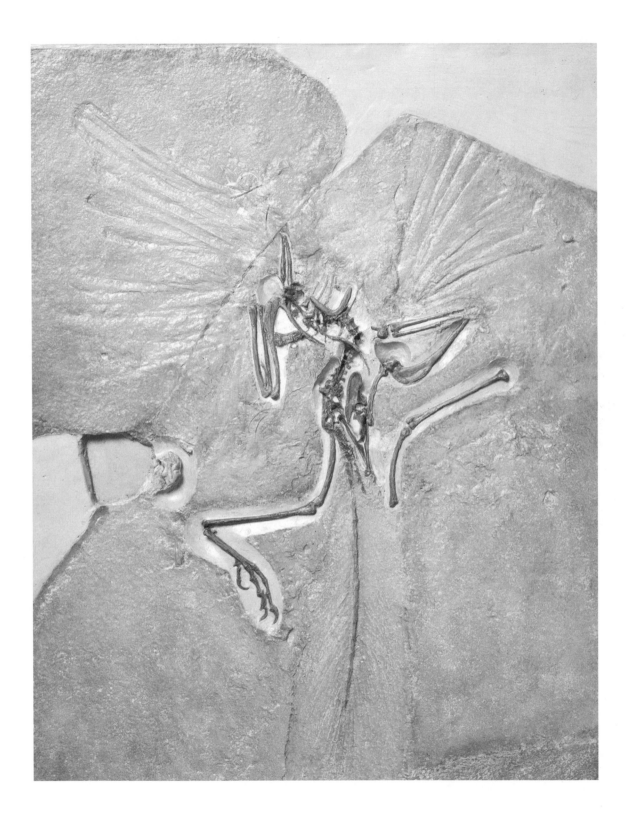

8

Archaeopteryx

It's a feather – about two and
a half inches long, a bit fluffy
at the base and in a couple of
places split where the
filaments on either side of the
central quill have separated.
An ordinary looking feather
really – except that it is a
hundred and fifty million
years old.

◁ This is the first skeleton of the bird-reptile to be discovered and recognised.
As the dying animal flopped into the salty tropical lagoon, its feathers and
outstretched wings must have slowed the rate at which it sank to the muddy
bottom. Few if any carrion feeders lived in these extremely salty waters.
So the skeleton remained articulated and remarkably complete as the silt
settled over it. Only its head is missing. A small fragment of the skull, however,
lay quite close by. Although later specimens were more complete,
this one, in the Natural History Museum in London, is of great importance
for it was this that gave the species its name.

It's a feather – about two and a half inches long, a bit fluffy at the base and in a couple of places split where the filaments on either side of the central quill have separated. An ordinary looking feather really – except that it is a hundred and fifty million years old.

It lies on a small piece of limestone and was found in a quarry near Solnhofen in Bavaria in 1860. The stone there has been worked since Roman times for roofing tiles and masonry – as it still is – but it is also famous for the perfect fossils that are occasionally found in it. The limestone is the solidified compressed mud that once accumulated on the bottom of a shallow tropical lagoon. It seems that the lagoon was not a healthy place for animals. The water was hot and very salty. So there were very few permanent residents. But animals of one sort and another were occasionally washed in by the tide from the sea beyond – so there are shrimps and crabs and fish, all preserved in perfect detail. The bodies of land-living animals were also occasionally washed down into it by rivers – beetles, lizards and small dinosaurs the size of chickens. And some things fell into it from the sky – pterodactyls, dragonflies – and this feather! But what did *that* come from? Only birds have feathers. What kind of bird was it that flew alongside pterodactyls?

The answer turned up a few months later. It was a very strange bird indeed. In fact it was as much reptile as bird. It had feathers on its forelimbs, just like a modern bird, so you could certainly describe them as wings – except that *these* wings had separate fingers each of which ended in a claw. And it had an extraordinary tail – bony, like a lizard's tail – except that there were feathers sprouting from either side. Only fragments of the head remained and they had become separated from the main skeleton. But one piece of the lower jaw showed that it had teeth. This amazing fossil came into the possession of a local doctor who looked after the quarrymen and who allowed his patients to pay him with fossils. He had built up a large collection in this way, but now his daughter was about to get married and he wanted to give her a decent dowry. So he put his whole collection up for sale – including the strange feathered creature which had by now been given a name – *Archaeopteryx*, 'ancient wing' in Greek.

Its availability came to the attention of Richard Owen who was in charge of natural history at the British Museum and who would soon build and become first director of the Natural History Museum in South Kensington. He determined that this important fossil should come to his new museum. There was intense competition with German museums and not a little dubious dealing but in the end Owen got it – for seven hundred pounds, which was an enormous sum in those days.

This was a critical moment in scientific history. Charles Darwin's revolutionary book, *The Origin of Species*, had been published only two years earlier in 1859. In it, Darwin claimed that animal groups were descended from one another. Owen on the other hand maintained that God had created, one by one, the basic pattern for a group – an archetype he called it – in much the same way as described in the Book of Genesis. The Almighty then created individual species by, as it were, tweaking that archetype, exaggerating a detail here, eliminating one there. And *Archaeopteryx*, said Owen, was clearly a bird.

There were heated arguments, but eventually most scientists accepted that Darwin was correct and *Archaeopteryx* was clearly an intermediate form – half-reptile half-bird. So Owen was left in the embarrassing position of holding in his museum one of the most convincing proofs of a theory he himself did not accept

And there matters rested, with *Archaeopteryx* the undoubted star of the Museum's fossil galleries, until over a century later. Then in 1986 – extraordinarily and sensationally – two distinguished scientists, astronomers not biologists – declared that the fossil was a fake. Fred Hoyle was a brash but brilliant astronomer whose radio talks on the origin of the universe, delivered in a strong Yorkshire accent, were extremely popular. He had decided, together with a colleague, Chandra Wickramasinghe, that the driving force of evolution was not natural selection as proposed by Darwin but events they called 'genetic storms' during which the earth was sprayed from outer space by genetic material rather like viruses.

The notion that there were intermediate forms linking the major groups did not suit this theory. Hoyle declared that whereas the bony parts in the *Archaeopteryx* fossil were perfectly genuine, and were, in fact, the bones of a dinosaur, the impression of feathers were faked. This had been done, he said, by grinding up some Solnhofen limestone into a paste, smearing it around the dinosaur bones and then pressing chicken feathers or some such into the paste to create the feather impressions. The accusation was particularly absurd, since by now four other *Archaeopteryx* fossils had been found, complete with feather impressions, including perhaps the finest of all, complete with a skull and jaws lined with teeth, which is now in Berlin.

Nonetheless, the accusations gained so much publicity that the Natural History Museum had to take them seriously and the resident dinosaur expert there at the time, Alan Charig, had to stop his dinosaur research and turn his attention for a year or more to examining the fossil in the greatest detail. The Museum owns not only the slab that contains the majority of the fossilised bones, but also the one that lay directly on top of it – the counterpart, as it is called. Charig showed that not only do the two slabs fit together perfectly, leaving no room for any additional paste, but that mineral patterns on the feather impressions on it also appear on the counterpart – and the two match perfectly. His arguments were conclusive and that was the end of that.

Since then, new techniques have made it possible to extract even more information from this wonderful fossil. The Museum scientist who currently takes care of *Archaeopteryx*, Angela Milner, has managed to reconstruct the partially damaged braincase and examine the contours of its interior using high resolution computed tomography – the CT scanning technique that is now used widely in hospitals to look into our bodies. *Archaeopteryx*'s brain would have fitted so closely against its skull that the internal contours of the brain case make it possible to deduce the shape of the brain itself. That told Angela Milner that *Archaeopteryx*, when alive, had both a well-developed sense of balance and an acute and sensitive vision – both characteristics that are necessary for efficient flight and that are possessed by modern birds.

And now a whole new set of fossils have been discovered in China that have added more links to the reptile-to-bird chain. The earliest clearly belong to a family of small, flesh-eating dinosaurs known as theropods – and they have coats of feathers clearly visible as a brown halo on the limestone around their skeletons. But the skeletons also make it clear that these very ancient creatures could not fly. Why then did they have feathers? The answer seems to be that they were able to generate heat within their bodies and that the feathers provided a warm coat that prevented that heat from being wasted. They were able, therefore, to be active early in the morning and go out to hunt for food before the sun warmed their cold-blooded competitors – a crucial advantage.

There are two views as to how these reptiles in their feather coats started to colonise the air. One is that they started to run on their hind legs with their feather-covered forelegs outstretched. This may have enabled them, according to this theory, to trap insects –

or maybe to lift themselves into the air just before their pursuers were able to grab them. That doesn't seem likely to me. For one thing, if you need to escape from predators, you can run faster on four legs than you can on two. Even an Olympic sprinter can't keep up with a cheetah. And if, as you run for your life, you rear up and stretch out your feather-clad forelegs, that will slow you down even more.

A more likely explanation, to my mind, is that these little well-wrapped dinosaurs – if we can still call them that – started to clamber around in the trees hunting for insects and other climbing reptiles. Spreading their forearms enabled them to glide from branch to branch and tree to tree. But either way, *Archaeopteryx* remains the crucial link. For me, it is one of the great wonders in the Natural History Museum – surely the most convincing proof of evolution that you could possibly get on a couple of square feet of roofing tile.

This single feather, discovered in 1860, was the first indication that there were feathered creatures in the sky flying with the pterodactyls and dragonflies above the Solnhofen lagoon.

The most complete and perhaps, with its almost symmetrical pose, the most beautiful of all *Archaeopteryx* specimens so far discovered, was found sixteen years after the first London specimen. It is now in Berlin Museum. It shows that although the animal had well-feathered wings like a bird, it had, instead of a beak, jaws lined with teeth, like a reptile.

Richard Owen, newly appointed as superintendent of the natural history department of the British Museum, recognised the importance of the *Archaeopteryx* fossil and managed to raise the huge sum of money needed to acquire it for his institution. He could not accept, however, that it was spectacular proof of the reality of organic evolution as explained by Charles Darwin only two years earlier.

Fred Hoyle, a brilliant astronomer and very successful populariser of his science believed that life originated in outer space. He did not accept that animals might evolve and perhaps because *Archaeopteryx* demonstrated otherwise, he maintained that the impression of its feathers had been had been faked. It took Museum scientists a year or so to prove conclusively that he was wrong.

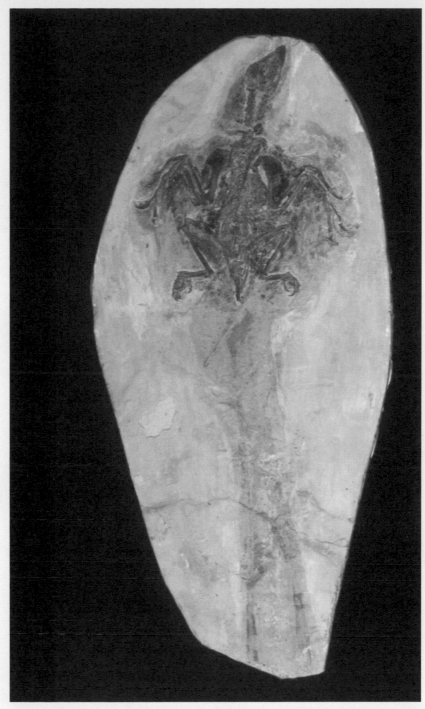

In 1995 extraordinary fossils were discovered in the Chinese province of Liaoning. dating from some thirty million years after *Archaeopteryx*. Among them were feathered animals intermediate between *Archaeopteryx* and modern birds. This species was called *Confuciusornis* in honour of the great Chinese philosopher. Its bony tail has been reduced to a mere stump and, most significantly, in place of a heavy bony jaw containing teeth it has a horny beak.

9

salamanders

My first pet was a
salamander. 'Pet', perhaps, is
not the right word, if it brings
to mind something you can
stroke, a being that will
respond to you and ultimately
become your companion.
Salamanders, I have to admit,
are not really for cuddling.

◁ European salamanders vary greatly in their coloration. Some individuals may
be almost entirely black with small yellow spots, others the reverse. The top three
come from mainland Europe. The middle one is from Corsica and the lower three
are from the Alps.

M y first pet was a salamander. 'Pet', perhaps, is not the right word, if it brings to mind something you can stroke, a being that will respond to you and ultimately become your companion. Salamanders, I have to admit, are not really for cuddling.

They are amphibians, and, like all such creatures, they are cold-blooded with moist skins like a frog. Worldwide there are several hundred species of them. In shape they look like rather corpulent newts, but they do come in a variety of sizes. The biggest are over six feet long – brown, fearsome-looking monsters with blunt heads and tiny bead-like eyes. They live in Japan and China and spend nearly all of their time in the water. The smallest are slim creatures only an inch and a half in length that wriggle through the leaf litter on the floor of the forests of eastern North America. No one sees much of them for their bodies are very long, almost worm-like with very small legs and they spend nearly all their time deep in the leaf litter on the forest floor. My salamander was an intermediate sort, about seven inches long and from Europe. It was a fire salamander and my father gave it to me as a present to mark my eighth birthday.

It didn't move very fast so I was able to pick it up and allow it to clamber around on my hand. Its cold, clammy body felt rather rubbery and smelt faintly of vanilla, but the most remarkable thing about it, to my young eyes, was its colour. It was a dramatic glossy black with brilliant sulphur yellow blotches. It seemed as glamorous and as spectacular to me as any tropical species.

No one seems to know how it got its popular name of fire salamander. Some say it is because of an ancient legend that claims it has the ability to quench fire. Maybe its connection with fire is because the animal itself is the very opposite. Lizards, which have much the same shaped bodies, love heat and bask in the sun. The salamander, on the other hand, very surprisingly much prefers cold damp places. Pliny, the down-to-earth Roman naturalist knew the legend well and decided to test it rather callously by putting a salamander in a fire. Needless to say the poor thing was burnt to a cinder. He gave a full account of this in his book but nevertheless the legend – and certainly the name – has persisted.

Salamander colouring – black and yellow – is a standard signal in the natural world for poison – danger. Wasps have it. So do some venomous snakes such as the banded krait, and the Gila Monster, one of the very few poisonous lizards in the world. The fire salamander has it, presumably as a warning of the fact that it possesses two glands on either side of its head from which, if it is provoked, it can secrete a milky fluid that is mildly poisonous.

The meaning of its colouring was not lost on people in the Middle Ages and the salamander was greatly feared. One writer reported as sober historical fact that four thousand men and four thousand horses of the army of Alexander the Great all died, simply because they had drunk from a stream through which a salamander had recently passed. Even Edward Topsell, who compiled an English animal encyclopedia in the seventeeth century, maintained that if anyone touched what he called 'the spittle' of a salamander they would lose all their hair.

My salamander, however, seemed perfectly innocuous and would allow itself to be picked up and inspected without any sign of getting irritated or secreting anything whatsoever. I kept it in a glass tank, furnished with moss and a large piece of cork bark under which it could shelter. At one end there was a little pool of water, floored with gravel in case it wanted a swim although, as I discovered later, a fire salamander is really very much more at home on land than in water.

I have to admit that it didn't *do* much – apart from rather reluctantly snapping up the slugs and worms that I provided for it. Nonetheless, somehow or other, it managed to escape fairly regularly, usually at night after I had gone to bed. Doubtless that was because, after my daily inspection, I had been careless in replacing the glass cover of the tank. The first time it vanished, I was distraught. Frantically, I searched the garden and at last discovered it beneath a cat-mint bush. Thereafter, whenever it disappeared from its tank – which was not infrequent – I knew exactly where to find it.

It made such an impression on me that when my own son reached *his* eighth birthday, I gave *him* a fire salamander. Having been warned to be careful, he unpacked the box in which it came, gently removed the wet moss around it and – very gratifyingly – he seemed just as astonished and thrilled as I had been when I was his age. Together we prepared a glass tank in just the same way as I had done all those years ago and gently put the salamander in its new home. It explored it, in a rather plodding salamanderish kind of way and then waddled down to the little pool at the end. There, it turned round so that its rear half was in the water and – to the total astonishment of both my son and myself – miniature salamanders, barely an inch long and complete with tiny yellow spots on their black skins, began to appear from the vent on the female's underside. One by one, they clambered out of the water and hid in the moss. In the end, there were over a dozen of them.

I had some excuse for my astonishment. The reproductive habits of the European salamander *do* vary. Most, in the south, produce tadpoles. Initially these have external gills but no legs and they live in water until they change into the adult form. Only a few adult females – mostly those that come from the northern part of their European range – retain their eggs and allow the hatchlings to complete their development actually within their mother's body.

So at least my gift had helped me in my parental duty of explaining to my son where babies come from – from inside their mother. But it did not help in the other part of that important responsibility – of explaining how they got there.

Salamander reproduction is, in fact, a very strange business. A male starts by depositing a blob of sperm on the ground. He then crawls beneath the female, entwines his front legs with hers and heaves her on to his back, like a porter picking up a heavy pack. He takes her to where he has deposited his sperm and then, with a twist of his hips, tips her half off so that her vent is directly over it and she can take it in. That process must have happened some weeks before our female had come to us and we never did witness it.

Rearing our salamander babies was not difficult – though rather tedious. We presented them with tiny fragments of meat which we held in forceps and waved about in front of their noses. They eventually got the message. But salamanders, I must admit, are not really very quick on the uptake. Looking after them was therefore a fairly time-consuming job even when we had weaned them on to worms, slugs and woodlice, for they still preferred their meals to be presented in front of them rather than search for them themselves.

Eventually we managed to spread the load. I managed to convince some of my son's school friends that they too ought to have a salamander. I also succeeded in passing on a few to various nephews and nieces. Eventually we found homes for them all, except for a couple that we kept ourselves.

As is the way of things, children do sometimes lose interest in their pets – and then their mothers have to take them over. So for many years afterwards I occasionally met a lady among my more distant neighbours, who would blame me for the fact that even though her children had left home years before, she was still condemned to scour her gardens nightly for slugs and then to waggle them with forceps in the front of elderly salamanders that had been born in *my* house. I, of course, apologise. I never dare to tell her that, in captivity, fire salamanders have been known to live for fifty years.

Legends about the lethal bite of a salamander were still being repeated in the seventeenth century by Edward Topsell in his *Historie of Four Footed Beasts* from which this illustration comes. In fact, as Topsell admits, it is very difficult to persuade a salamander to bite anything much bigger than a worm.

Black and yellow signal danger and many highly poisonous species, like this Banded Krait from India, warn others to keep away from them by being coloured in this way.

Warning colours, however, can also be flaunted as a bluff by entirely harmless creatures. This little frog from Australia has no poison glands in its skin. Whether or not the mildly unpleasant poison secretions that a salamander can on occasion produce are sufficient for it to be avoided is questionable, but its coloration certainly convinced people for many centuries that it had lethal powers.

The long almost worm-like bodies of the North American salamanders are well suited to wriggling their way through the leaf litter where they spend most of their time. Such a shape also produces a high surface to volume ratio, and that enables them to absorb all the oxygen they need through their moist skins. They do this so efficiently that all the species in this very large salamander family have now lost their lungs.

The biggest living salamanders are the giants that live in the streams of China and Japan. They can grow to lengths of five feet and seldom leave the water.

Most amphibians breed in water. There, sperm simply drifts on to the eggs. On land, where the fire salamander spends most of its time, it is more difficult. The male therefore first deposits a blob of sperm on the ground and then hoists the female on his back and carries her to it.

10

Birds of Paradise

Birds of paradise – as far as I am concerned – are the most glamorous of all birds. There are forty-two different species in the family. Most have costumes of feathers that have an unlikely, almost ludicrous, extravagance.

◁ The Red Bird of Paradise which lives on one of New Guinea's large off-shore islands is, in general shape, very similar to the Lesser and Greater Birds of Paradise from the mainland and the Aru Islands that were the first to reach Europe. The main difference is that whereas they had golden yellow plumes, the male Red Bird has plumes that are a deep rich red. When he is at rest, these are largely concealed beneath his wings. Only their pale white ends are visible as two tufts on either side of his tail. Two of his tail feathers have become long naked quills that trail behind him.

Birds of paradise – as far as I am concerned – are the most glamorous of all birds. There are forty-two different species in the family. Most have costumes of feathers that have the unlikely, almost ludicrous, extravagance you see in the pages of fashion magazines.

You know the sort of thing – clothes that you can't believe could possibly hold together in real life or are so unwieldy that the model wearing them can barely stand upright, let alone walk. But most of the models – and certainly all of the birds – are wonderfully and incontrovertibly beautiful.

The first species of bird of paradise to reach western Europe had long golden plumes sprouting from beneath their wings. The filaments which in a normal feather zip together to form an air-catching surface on either side of the quill, in these plumes are greatly reduced in number and widely spaced so that the plumes become a hazy golden gauze of quite ethereal beauty. They arrived in Portugal on September 6th 1522, one of the marvels brought back by Magellan's ships from their two-year long voyage during which, for the first time, they had sailed right round the world. The King of Bachian, one of the Spice Islands, now in the east of today's Republic of Indonesia, had sent the specimens as a gift to the King of Spain. And very odd they were – merely skins from which the legs, the body and all the skeleton except for the beak, had been removed. The wings too had been cut off so that the flank plumes could be seen in their full glory. Trade in these extraordinary and wondrous objects had been going on for centuries all over Asia. In Siam and Nepal the plumes were thought so precious that they were – and indeed, they still are – part of the royal regalia.

The fact that these birds apparently lacked both wings and feet baffled Magellan's sailors when they were shown them in Bachian.

Without such things how could the birds perch or fly? The Bachian-ese, driven to produce some kind of explanation, maintained that the birds didn't need to do either because when they were alive they floated in the heavens, feeding on dew. People only found them when they died and fell to earth. That was why they called them *bolong diwata,* 'bird of the gods'.

The truth, of course, was that the Bachian-ese had no idea what the living birds were really like. None of them had ever seen one alive. The skins, already cropped of their wings and legs had been traded to Bachian from an island far away to the east.

Magellan's men, however, passed on the paradise story and the sixteenth century naturalists, who were still convinced that there were mermaids and fire-breathing dragons to be found in foreign parts, happily accepted it and named the birds in Latin, the language of scholarship in those days, *Paradisea apoda*, the footless birds of paradise – a name that, because of the strict rules of scientific nomenclature, is still used for the bird we otherwise know as the Greater Bird of Paradise.

As more skins arrived, the European scholar-naturalists, sitting in their libraries cataloguing the world's curiosities, began to speculate on the natural history of birds of paradise. Feeding on dew made some sort of sense – at a pinch. But did these amazing creatures lay eggs? And if so, where did they build their nests up in the skies? Somebody dreamed up an explanation which, in no time at all, appeared in printed encyclopaedias as fact. Well, of course, the female laid them on the back of the male as he hovered about among the clouds. And then she sat on them – and him.

Some kinds of these wonderful birds, in addition to plumes, had naked curling quills projecting from their tail. What were *these* for? The explanation of that was obvious. When the birds got tired of hovering they came down to rest in the trees. Since they didn't have legs and couldn't perch, they hooked themselves on to the branches and then hung there, suspended by their tail quills.

These fantasies, of course, were slowly abandoned as scientific rationality spread among European scholars. But even so, when Linnaeus, the great Swedish cataloguer of the natural world, came to allocate an official name to the species, he retained the name 'apoda' – legless – and the species still carries it to this day.

Nonetheless, the true character of the birds remained largely a mystery for a long time yet. European explorers didn't manage to penetrate very far into mainland New Guinea, the bird of paradise's homeland, until the early nineteenth century. Then in 1824 a Frenchman, René Lesson, reached the island and saw one flying in the forest. In the decades that followed, slowly, species by species, the full wonder of the family was revealed.

It is only the male birds that have the plumes. All the species are polygamous. As a consequence, the males take no part in nest building, in incubating the eggs or in rearing the young. Instead, they spend their time competing between themselves to be the one who is most desirable. The winner then mates with all the females in the neighbourhood. Sometimes they compete by assembling in the same tree alongside one another and then when a female appears, snapping into a frantic dance to show off their plumes. Sometimes they dance separately, each performing in his own display arena, distributed throughout the forest. The females then tour from one to another inspecting the candidates.

Of course, birds of paradise are not the only family of birds to use this courtship technique. In Asia, many species of pheasants do so. So do ruffs in East Anglia and blackcock in Scotland. But birds of paradise do it more spectacularly and in greater variety than in any other bird family. Why should that be so?

The answer has to do with the unique nature of the thousand-mile long island of their New Guinea home. It lies in the tropics, so it has high temperatures and heavy rain throughout the year – just what life needs to flourish in abundance. So it is covered in thick rain forest. But in geological terms it is comparatively young. It was pushed up from the bottom of the ocean by the irresistible northern drift of the island continent of Australia. It is so new, in fact, that comparatively few mammals have managed to reach it – and there are very few ground predators. So the birds, which of course flew there, have been able to able to flourish.

The ancestors of the bird of paradise family, which were probably crow-like birds, were therefore living in what could be accurately called – in bird terms – a paradise where they had very few enemies and so much food – insects, leaves or fruit according to choice – that a female was able to care for her young entirely by herself. The only help she needed from the male was to contribute to that swift moment of fertilisation. Thereafter she could manage to raise her family unaided. Released from parental duties, the males were then able to concentrate all their time when they were not feeding on impressing the females.

That, of course, is what courtship is mostly about. In many species of birds, the males demonstrate in practical terms what valuable mates they will be. Male eagles do so with displays of the aerobatics needed for hunting; male weaver birds by showing how skilled they are at nest building; male terns by presenting the female with a row of fish carried neatly cross-wise in the beak. But these practical skills don't particularly impress female birds of paradise. They don't require the male to build a nest or feed the young. So the males beguile the females by displaying their plumes and demonstrating that they are the most beautiful males around.

And the females respond, selecting the male, out of sometimes a dozen or so that are swanking about within her range, that seem to her to be the most handsome. Populations diverged when one group of females developed a taste for, let us say, red plumes, whereas others preferred sulphur yellow ones. Well, as we know, tastes can vary and in the world of fashion they seem to do so more extremely than anywhere else. So perhaps it should not come as a surprise that the bird of paradise family is about as diverse as you can imagine. One species has a head-dress like one of the dottier kind of Ascot hats, that consists of six vertical quills each of which ends in a little black pennant. Another has a pair of quills, sprouting from the back of its neck that stretch for twice the length of its body and are decorated with what look like enamelled diamond-shaped plates. Yet another has scarlet flanks a sulphur yellow back and a naked ultramarine scalp.

High fashion, I know, is inclined to be extreme and quirky – but I don't think you can get more extreme and quirky than that.

At the beginning of the 17th century naturalists still believed that birds of paradise lacked wings and feet and floated in paradise, high in the sky. Ulysses Aldrovandus showed one in his great encyclopaedia that was published in 1601, drifting among the clouds, clad only in its gossamer plumes.

Bird of paradise plumes are still used as currency among many of the people of New Guinea and at ceremonial gatherings men continue to wear them as head-dresses that indicate their wealth.

So beautiful and extraordinary are the paradise plumes that they have been traded and treasured far beyond the shores of New Guinea for many centuries. They still form part of the royal regalia of Nepal and Thailand. And the young Scottish prince who in 1625 was to become King Charles I of England, had a hat, here resting on the table beside him, decorated with such golden plumes, complete with the bird's beak, secured with what appears to be an emerald studded brooch.

The Red Birds of
Paradise, isolated on
their islands,
developed their own
special way of
impressing females.
Instead of erecting
their plumes above
their back in a
shimmering
fountain, as Lesser
and Greater Birds do
on the mainland,
they show off to a
female by hanging
upside down,
spreading their wings
and displaying their
plumes vertically.

The Twelve-wired Bird of Paradise not only has a bunches of yellow plumes projecting from each of his flanks, but six extremely long naked quills that are bent upwards. Not surprisingly, these quills mystified the artists who had the task of imagining, from the evidence of the dried shrivelled skins what the living bird must have looked like. This is the attempt made by of one of the great French bird artists, Jacques Barraband in 1806.

William Hart, one of the several artists employed by the naturalist publisher, John Gould, like Barraband never saw the living Twelve-wired bird and had to use skins to prepare his painting of it. His version approximates fairly closely to reality. The function of the naked quills – the wires – that are so oddly bent back, remained unknown until nearly the end of the twentieth century when it was at last observed that the male uses them to flick the female in the face during his courtship dance

SELEUCIDES NIGRICANS, (Shaw)

The Magnificent Bird of Paradise, which is about the size of a thrush, is a quick-change artist. Perched in the branches above his display ground, he calls to summon the females, exposing the lime green lining of his mouth. At the same time he spreads his iridescent emerald green breast-shield, expanding and contracting it rhythmically so that it seems to pulse.

When the female appears, he hops down to a lower perch, flattens his breast-shield and suddenly erects a sulphur yellow cape that he has at the back of his neck, which from the front has the iridescence of spun glass, so revealing the whole of his golden back. Leaning outwards, exposing his ultramarine feet, he projects the two curling tail quills so that they are parallel to the ground. This female, in her comparatively drab plumage, watches this extraordinary performance, it seems, with a highly critical eye.

This, the Ribbon-tailed Bird of Paradise, was the last to be discovered. In 1939 a naturalist spotted its spectacular plumes in the head-dress of a warrior who was living in part of the island that till then had hardly been explored by outsiders. The male's white tail plumes may be three feet long. It displays by flicking them from side to side and jumping between the branches of a tree.

The male King of Saxony Bird of Paradise has two extraordinary quills over twice the length of his body growing from the back of his head. Each carries a line of up to fifty small pale blue flags, that look like fragments of enamel. To display, he perches on a long vine looped down between the branch of a tree. He nods his head so that his two head plumes thrash backwards and forward, and at the same time rhythmically flexes his legs with such vigour the whole vine bounces as if he were on a trapeze.

Carola's Parotia Bird of Paradise performs his displays on the forest floor. He clears a patch of ground of all twigs and fallen leaves, and uses it as a stage. To announce his performance he calls loudly. When at last he has attracted a suitable audience – here five females – he erects his flank plumes so that they form a circular crinoline and begins to dance. With tiny rapid steps he sways from side to side while the females watch. If one of the females is sufficiently impressed, she will stay on the perch when at the climax of his dance he suddenly flutters into the air and lands on her back to mate.

11

The serpent's stare

There is something profoundly unsettling about snakes. It's the way they look at you with that steady unblinking stare. It's as though they have come from another world. And indeed, ancestrally, they almost certainly have.

◁ A snake, the agile and venomous Boomslang of central and southern Africa, stares with lidless eyes – but its sight is not good.

There is something profoundly unsettling about snakes. It's the way they look at you with that steady unblinking stare. It's as though they have come from another world. And indeed, ancestrally, they almost certainly have. They have come from the underworld.

There is a lot to be said for a life spent beneath the surface of the earth. Conditions are steadier than on the surface. There is no wind to buffet you, and no great swing of temperature between night and day. You are out of the way of other larger dangerous animals that might be seeking to make a meal of you. And there is a surprising amount of food – roots and bulbs if you are a vegetarian, and if you are not, worms, beetle grubs, termites and other insects, as well as, of course, others of your own kind that have taken to the subterranean life.

Backboned animals first clambered up out of the water and on to the land around a hundred and seventy million years ago. But it seems that quite soon after that, some of them started to burrow and disappear below the surface of the earth. One of the earliest to do so was a branch of those very first four-legged back-boned animals, the amphibians. They are called the caecilians.

The name is hardly familiar – and you are not likely ever to have seen one of these extraordinary creatures. Yet they are surprisingly abundant, particularly in the warmer parts of the world. There are a hundred and sixty odd species. Most are around a foot in length but one, from Brazil, is almost five feet long and is as thick as a man's arm. They have concertina-like grooves encircling their hairless moist bodies but these are superficial, a by-product of the way the animal burrows through the ground, pushing its pointed head through the soil so that its body is extended and then hitching itself up so that the concertina-rings close. There are no traces of legs, either externally or even internally. But if you are incautious enough to pick one up, you know immediately that it is not a giant earthworm, for its long tubular body does not flop about. This animal *writhes*. It has a backbone. And what is more, it bites – with small but very sharp teeth.

Below ground, of course, it is impossible to *see* anything. So eyes are of no use. Consequently the caecilians' eyes have not only degenerated into useless relics, but are covered in skin and have no function. The animals do, however, have a special sense organ all of their own. It's unlike anything else developed by any other animal – a pair of little tentacles on either side of the head, each lying in a groove between the corners of the mouth and the remnants of the eye. The caecilian can erect them and wave them about; and since there are nerves connected to them, it seems certain that they are sense organs of some kind – to help the animal to find its prey and, one presumes, its mate – by smell.

The most recent backboned animals to abandon light and air and go below are mammals – moles. They went down around some 45 million years ago, so recently in evolutionary terms that they have not yet lost their arms and legs as the caecilians have. The eyes of European moles, although tiny and buried deep in the fur, are still sufficiently functional to enable their owner to tell the difference between light and darkness. And there are other moles in Africa whose eyes are covered with skin so that they are totally blind and have a head that is reduced to little more than a furry wedge with a naked leathery front edge.

The most important group of backboned animals to have made the move underground are the reptiles. Different families have done so at different times. One, the amphisboenids or blind snakes, did so quite early on in reptilian history and are still there. Others, such as the skinks, still seem to be in the process of doing so. Skinks are those stout lizards with scales that are so smooth and fit so neatly together that the animals seem to be polished and are a delight to hold. Between them, they illustrate almost every stage of leg loss. Some have four legs that they can waggle but which are virtually useless for getting around. Others have lost their hind legs. And some called sandfish which live in dunes, have lost all their legs and swim, as it were, through the sand with wriggles of the body, moving at such speed that they are very difficult to catch.

But to come back to snakes. It is clear from many examples that animals living underground tend ultimately to lose both their limbs and their eyes and ears. And snakes certainly lack legs – superficially at least. But some do have relics of them. Boas and pythons have two small claw-like spurs either side of the vent which are each connected internally to one or more elongated bones deep in the flesh.

Externally, they have no sign of ears or of a tube leading to an eardrum. Instead their inner ear is connected by bones to the jaw so that snakes hear – if 'hear' is the right word – by resting their jaws on the ground and detecting vibrations.

But snakes *do* have eyes. Surely if they, in the past, lived underground, they would have lost them. Well, their eyes, in fact, are extremely odd and quite unlike the eyes of any other land animal. For one thing, they lack lids, so snakes can't blink. Instead, their eyes are protected by large transparent scales that come away when the animal sheds its skin. And a snake cannot move its eye in its socket as well as we can, for whereas we have eight muscles to do that, a snake has only six. As for focus, we and most other backboned animals achieve that by altering the curvature of the lens within the eye. Snakes do not have that ability. Instead they have developed a way of moving the whole lens backwards and forwards. It seems that the snake's eyes have re-evolved – have been re-built, as it were – using a greatly reduced number of parts – further evidence that snakes did, in the past, live underground.

Fossil remains of truly ancient snakes are rare but there is a snake-like skull that dates back to around 85 million years ago. That was the time when the dinosaurs were rampaging around the land, when underground was, by and large, a good place to be.

But then, of course, after many millions of years the dinosaurs suddenly disappeared. When they did, it became relatively safe for snakes to return to the surface. There were not many big birds, at that stage in evolutionary history, to threaten them. But there were quite a lot of nice little furry mammals to eat. Once above ground, the snakes began to prove that running on four legs is not after all the only effective way for a backboned animal to get around. They use several methods but most can curve their bodies into an S-shape so that their flanks can get a purchase on irregularities on the ground. And if you have ever tried to chase a snake, you will know that many species can move across the ground much faster than you can.

Some snakes have also developed special additional sense organs. They smell with their fork-shaped tongues – a talent they may have inherited from the land-living group from which they descended, the monitor lizards. By flicking the tongue in and out, they are able to collect scent molecules from the air and carry them back to a special sense organ in the roof of the mouth. But the most highly developed snakes, vipers like rattlesnakes, have a pair of small but deep and narrow pits in front of the eyes. These are sense organs like no other in the animal kingdom. They are heat detectors, so sensitive that they can register the presence of a small warm-blooded creature such as a mouse up to eighteen inches away. What is more, that sense is narrowly focussed into what amounts to a beam. And since the snake has two such pits, those beams enable it to locate a mouse so well that the snake's strike, when it comes, hits its target with great precision, like the punch of a skilled boxer, neither falling short nor overshooting. And since the rattlesnake is primarily a night-time hunter, those heat-detection pits are much more valuable than eyes.

But even so, the unblinking but vague stare of a snake still un-nerves me – and maybe you. These creatures are aliens. They almost certainly have come from another world – even if it is a long time since they were there.

The cockatrice was believed to hatch from a spherical soft-shelled egg laid by a seven-year-old cockerel and incubated by a toad. Superficially snake-like and only a few inches long it was one of the most deadly of all creatures. Its very glance was lethal. Thus one way to deal with it was to use a mirror to reflect its gaze and so make it kill itself with its own power.

Both of these animals have a moist skin and live underground. The one being eaten is a worm, an animal without a backbone; the other has a backbone as you would soon discover if you picked it up. It also – visibly – has a powerful bite. It is a caecilian, a Brazilian amphibian that has taken to the underground life. The rings round the worm's body reflect the fundamentally segmented character of the animal's body. Those on the caecilian, however, are merely skin deep and are associated with the concertina-like way the animal moves through the soil.

This burrower also has a backbone, but unlike a caecilian, its skin is dry and scaly. It is a reptile, an amphisboenid. The name comes from Greek words meaning 'going both ways' for its head superficially is somewhat like its tail.

But it is a snake that lives entirely underground and has lost not only its legs but also its eyes.

Mammals also took to the underground life. The European mole still retains its eyes though they are very small. Fur below ground in not very practical. Mole rats have lost it altogether, but the mole still retains it. The hairs however, are very short so they can be brushed in any direction, which makes their fur very soft to the touch.

Golden moles are so called because their fur has a glossy yellowish sheen. Although some do look somewhat golden, most species – of which there are twenty – are brown or grey. They burrow their way through the dry sand at considerable speed by hunching their extremely powerful shoulders. Their eyes are mere vestiges and covered with skin, their ears are buried deep in the fur, and their nostrils are protected by a leathery wedge at the front of the head. They are not closely related to the European mole. The similarity between the two families is due to their similar way of life.

The skink family of lizards seem to be in the process of taking to an underground existence. Species can be found to illustrate almost every stage of leg-loss. This species has legs that are so small that they are of little help in moving around and the skink burrows through the leaf litter or sand with serpentine wriggles of its body.

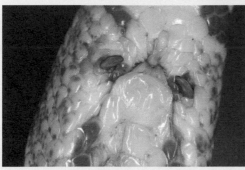

The spurs on either side of the vent of a python are used during mating. Those in the male are considerably bigger than in the female. They are, however, the relics of limbs that ancestral snakes once possessed. Internally there are small bones on either side that are the much-reduced remains of the pelvis and the thigh bones. The spurs are the remains of the claws.

A rattlesnake has advanced detector systems that more than compensate for its limited sight. Its flicking tongue gathers the molecules from the air that we perceive as smells and evaluates them on a sense organ in the roof of its mouth. Beneath each eye, there is a pit which is sensitive to the slightest change in temperature and can sense the warmth of a rodent's body a yard away. Its eyesight may be poor – but then it hunts in the dark when even the best eyes are of little value.

12

Fossils and Fakes

I suppose it's true to say that it was one of the key moments of my life. The trouble is, I can't remember when it was. But I can, nonetheless, describe it in detail because I have been repeating that moment, off and on, throughout my life and the thrill has still not worn off.

◁ Trilobites are not closely related to any animal alive today. They appear in some of the most ancient of sedimentary rocks dating back 545 million years. They were at their most abundant about a hundred million years later, and then finally disappeared around 260 million years ago. They had hard shells as shrimps do today and regularly shed them. These empty shells fossilise as well as their complete bodies and are somewhat more abundant.

I suppose it's true to say that it was one of the key moments of my life. The trouble is, I can't remember when it was. But I can, nonetheless, describe it in detail because I have been repeating that moment, off and on, throughout my life and the thrill has still not worn off.

It was the moment when I first hit a lump of stone with a hammer so that it split apart – and there, perfect in every detail, glinting as though it had just been polished, was a coiled seashell, three or four inches across – an object of breathtaking beauty – and my eyes were the first to see it since its occupant died two hundred million years ago.

It was, of course, an ammonite, an extinct creature related to the nautilus that is still alive today, and that sailed at depth through the Jurassic seas. The limestone in which it lay was the solidified, compressed mud that had accumulated at the bottom of those seas. Today Jurassic limestone outcrops over much of the eastern half of Leicestershire where I used to cycle when I was a boy searching for fossils, particularly ammonites.

The rock that contains these fossils is particularly rich in iron and at that time it was still used as iron ore – so many of the quarries were still being worked. There were also others – the best from my point of view – that were overgrown and disused. They were my treasure fields. There were lots of different ammonites to be found, all with grand sounding names – *Dactylioceras, Phylloceras, Oxynoticeras*. One even carried the name of a nearby village – *Tiltoniceras* – because the first ever specimens of that species were collected from a quarry near Tilton and what is more I had found specimens of it there myself. In my eyes they all had a wondrous, almost miraculous quality.

The local people thought they were *literally* miraculous. They called them 'snake stones'. The name, I think, originated farther north in Yorkshire around Whitby. There, they said that St Hilda, the daughter of a local 7th century king, had founded two monasteries, one for men, one for women. Unfortunately, the grounds of these monasteries were infested by snakes. However, St Hilda had a solution. She prayed and turned the lot of them to stone.

A snag in this explanation, of course, was the fact that the coils of an ammonite do not have at the front anything remotely resembling the head of a snake. So local people obligingly satisfied any of the devout who might be worried by this deficiency, by carving one out of the limestone that surrounded each ammonite. It could then be sold for a good price to people who wanted to carry it as a charm that would protect them against disease.

The practice went on certainly until the late nineteenth century and even in the 1930's when I was picking up my ammonites, there were villagers who seemed to prefer the St Hilda explanation to the rather more amazing one that I tried to provide – that the rolling hills of Leicestershire were made of consolidated mud that had accumulated at the bottom of a sea two hundred million years ago.

People still carve additions to fossils to make them more valuable – and not to the superstitious but to the scientifically gullible. Certainly if you are prepared to give good money for a fossil, you ought to be prepared for the fact that some specimens have been – how shall I put it – improved.

A few years ago, I was filming in the Atlas Mountains of Morocco. Every now and then, we would pass someone sitting by the roadside with a row of fossils in front of them offering them for sale. And these – at the risk of sounding somewhat faithless to my first-love, the ammonites – were fossils that were even more spectacular and more miraculous. These were trilobites. Trilobites are considerably more detailed and more complex than my ammonites – and even older – around 400 million years old. They are, of course, long extinct. Indeed they were already extinct when the ammonites first appeared. They look superficially like giant wood-lice, with their bodies covered with segmented armour plates, and underneath numerous jointed legs like those of a shrimp, though you can only see that in unusual specimens that are exceptionally well preserved.

There are lots of different kinds, but some of those found in Morocco are huge – around five inches long. More miraculous still, they have a pair of large eyes – two conical projections, like little towers, on the front of the head, each covered with a mosaic of several hundred tiny pits, somewhat like those that make up the mosaic eye of a fly. The preservation in many instances is mind-blowingly perfect. It is so good, in fact, that their structure is preserved in microscopic detail. Each element contains a lens made from the crystalline form of calcium carbonate known as calcite.

Calcite is transparent all right, but rays of light passing through it are diffracted – split – unless, that is, they strike the surface of the crystal at one very precise angle. Then they pass through it undistorted. Amazingly, the calcite forming the lenses in the trilobite eye is orientated in exactly that way so that the image they transmit is in sharp focus. How can we be sure? Because one enterprising scientist has managed to take a photograph using one of them as a lens. These are the first eyes on the planet that ever saw an image of anything.

The magnificent fossils by the Morocco roadside were, of course, irresistible. But still, it seemed a bit too easy – just to pay my money and take away a fossil. Searching for fossils, for me, is still part of the fun. So I decided to try and find where they came from. Starting conversations with the sellers made it necessary to buy a trilobite or two, of course, but eventually we were able to follow the trail off the main road and along a stony bumpy track to a remote village up in the mountains. There, at last we found the mud-walled house belonging to the man from whom all the trilobites came.

He had hundreds of them in a great pile in his back room. I started to try and sort through them, but the room was only lit by a small glass-less window high in the wall. I was well aware that incomplete specimens are often made more saleable by carving the rock around them to look like the missing bits. Sometimes a complete front half is attached to the back half of another quite different one. But it wasn't all that easy to be sure one wasn't being taken in – and the piece of sacking covering the window kept slipping down so that it was really quite dark and difficult to see.

The owner kept producing specimens he called 'verry special – verry good!' Still, I couldn't make up my mind. But it was getting late and I had to be back at base that evening. He pulled my sleeve. 'This one', he said, 'Very *very* special, very very *rare*. Two! Together! Making love!!' And there they were in his hand, a pair, coiled round one another. I was astounded. There was not time to bargain and we really had to go. Could I resist a pair of love-making trilobites? Of course not. I paid him his price, rushed back to the car with my prize and then set off back.

We hadn't gone more than a mile or so out of the village when it dawned on me. Copulating trilobites? Surely not. I couldn't recall any learned scientific journal going into the mating habits of trilobites, but surely they must have reproduced at the bottom of the ancient seas by the female releasing her eggs on the sea floor and then the male simply releasing sperm over them to fertilise them. There would be no need in those distant days for them to get together. I must have bought a fake.

Late though we were, we stopped the car. I unwrapped the putative lovers and looked at them in the light of a torch. Of course, they were two separate specimens. One that was partly curled had been stuck to the underside of another with some rather unconvincing plaster. It was, I suppose, a lesson. But I still have the pair of them. They are tucked away in my cellar so that I am not continually reminded of my gullibility. But whether they were copulating or not, no one could reject creatures with eyes like theirs.

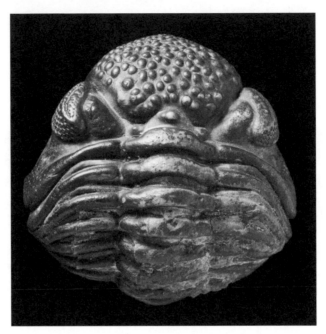

Trilobites could roll up to protect themselves, just as woodlice do today.

The dead body of an animal, particularly a shelly one like an ammonite, acts as a nucleus around which lime and other rock forming minerals tend to accumulate. So large hard nodules in a soft limestone or mudstone often reveal treasures for those fossil hunters who split them. This one, in two halves, contains an ammonite and was picked up on the Dorset coast.

The ammonites that are so
abundant in many rocks
seem clearly to be some
kind of mollusc, like a
snail. The fact that they are
surrounded by stone was
once a great puzzle. Some
people sought to explain
the mystery by invoking an
even greater one. They
suggested that these were
not any kind of snail but
snakes that had been
turned to stone by a saint.
And to give greater
credibility to the story –
and in order to sell the
fossils as holy relics – they
provided them with a
carved snake's head.

This is the largest
ammonite ever found. It is
1.7 metres – nearly 6 feet
across – and was
discovered in the chalk
rocks near Munster in
Germany

The rocks of Morocco are rich in minerals and fossils of many kinds. Visitors are offered both by shops like this one in the towns and also by small wayside stalls away in the mountains. And the people who sell them are very expert in preparing them so that they are made to look their very best – or even better!

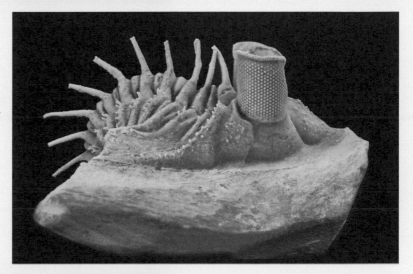

Towards the end of their era, the trilobites evolved into a great variety of extraordinary almost grotesque forms with all kinds of spines and other appendages.

The preservation of many of the Moroccan trilobites is extraordinarily detailed. This towering structure, in reality about ¼ inch high, is one of a pair of mosaic eyes that between them gave its owner 360 degree vision over the sea floor across which it crawled. Each element contains a lens of calcite. There is a curved rim at the top which shades the eye, so preventing light from above from dazzling the animal.

PREMIO   £ 100   REWARD
RÉCOMPENSE

Examine este peixe com cuidado. Talvez lhe dê sorte. Repare nos dois rabos que possui e nas suas estranhas barbatanas. O único exemplar que a ciência encontrou tinha, de comprimento, 160 centímetros. Mas já houve quem visse outros. Se tiver a sorte de apanhar ou encontrar algum NÃO O CORTE NEM O LIMPE DE QUALQUER MODO — conduza-o imediatamente, inteiro, a um frigorífico ou peça a pessoa competente que dele se ocupe. Solicite, ao mesmo tempo, a essa pessoa, que avise imediatamente, por meio de telgrama o professor J. L. B. Smith, da Rhodes University, Grahamstown, União Sul-Africana.

Os dois primeiros especimes serão pagos à razão de 10.000$, cada, sendo o pagamento garantido pela Rhodes University e pelo South African Council for Scientific and Industrial Research. Se conseguir obter mais de dois, conserve-os todos, visto terem grande valor, para fins científicos, e as suas canseiras serão bem recompensadas.

COELACANTH

Look carefully at this fish. It may bring you good fortune. Note the peculiar double tail, and the fins. The only one ever saved for science was 5 ft (160 cm.) long. Others have been seen. If you have the good fortune to catch or find one DO NOT CUT OR CLEAN IT ANY WAY but get it whole at once to a cold storage or to some responsible official who can care for it, and ask him to notify Professor J. L. B. Smith of Rhodes University Grahamstown, Union of S. A., immediately by telegraph. For the first 2 specimens £ 100 (10.000 Esc.) each will be paid, guaranteed by Rhodes University and by the South African Council for Scientific and Industrial Research. If you get more than 2, save them all, as every one is valuable for scientific purposes and you will be well paid.

Veuillez remarquer avec attention ce poisson. Il pourra vous apporter bonne chance, peut être. Regardez les deux queuex qu'il possède et ses étranges nageoires. Le seul exemplaire que la science a trouvé avait, de longueur, 160 centimètres. Cependant d'autres ont trouvés quelques exemplaires en plus.

Si jamais vous avez la chance d'en trouver un NE LE DÉCOUPEZ PAS NI NE LE NETTOYEZ D'AUCUNE FAÇON, conduisez-le immédiatement, tout entier, a un frigorifique ou glacière en demandat a une personne competante de s'en occuper. Simultanement veuillez prier a cette personne de faire part telegraphiquement à Mr. le Professeus J. L. B. Smith, de la Rhodes University, Grahamstown, Union Sud-Africaine.

Le deux premiers exemplaires seront payés à la raison de £ 100 chaque dont le payment est garanti par la Rhodes University et par le South African Council for Scientific and Industrial Research.

Si, jamais il vous est possible d'en obtenir plus de deux, nous vous serions très grés de les conserver vu qu'ils sont d'une très grande valeur pour fins scientifiques, et, neanmoins les fatigues pour obtantion seront bien recompensées.

13

coelacanth

December 1952.
An ancient fish suddenly
becomes headline news.
Indeed, it has become the
cause of an international
incident. France is at
loggerheads with South Africa,
claiming that in the dash to
grab the fish, South Africa has
violated her territorial rights.

◁ Professor J.B.L. Smith, having recognised the huge scientific interest in a living
coelacanth, prepared this reward poster which he sent to ports and villages all over
the western Indian Ocean from Madagascar and the Comoro Islands to
Mozambique and the shores of the Red Sea.

December 1952. An ancient fish suddenly becomes headline news. Indeed, it has become the cause of an international incident. France is at loggerheads with South Africa, claiming that in the dash to grab the fish, South Africa has violated her territorial rights.

I, at the time, am a young television producer working for the one and only television organisation in Britain, the BBC, which was producing all its programmes, live, from two tiny studios in Alexandra Palace up in north London. I am summoned by my boss. 'You' she said, 'were educated as a biologist – and you have been here for nearly two months now. So put on a programme next week to explain what it's all about. Fifteen minutes should do it'. That's how television was in those days.

There was one obvious thing to do – get in touch with Professor Julian Huxley. He was a scientist famous for his appearances on radio's Brains Trust. Professor Huxley agreed. He would need a few pickled fish – a shark, a lungfish, perhaps a salamander – and a photograph of the fish that was causing all the trouble. I rushed around, gathered such exhibits together and, in due course, the Professor appeared to explain to television viewers what all the fuss was about.

There was a time, around 375 million years ago, when the only animals on land were creatures without backbones – spiders, scorpions, gigantic millipedes and huge dragonflies with wings nearly two feet across. The only animals with backbones were fish. One of these, around this time, began to clamber out of the water and up on to land. That fish was the ancestor of all backboned animals that have appeared since – amphibians, reptiles, birds, mammals and, of course, ourselves.

But what fish was it? Whatever it was, it had to have strong fleshy fins, strengthened with bones, which would have enabled it to move around on land, unsupported by water. One strong candidate, well-known from its fossils, was called a coelacanth. The fossils showed its bones but they could not reveal anything about its internal anatomy, its soft parts, that enabled it to survive out of water breathing air. Then, back in 1938, a living coelacanth was hauled up by a fisherman working off the coast of South Africa. It had the typical coelacanth shape, with fleshy bases to its fins. But whereas the fossil coelacanths were only a few inches long, this was a monster, five feet in length. It was some time before South Africa's greatest fish expert, Professor J.B.L. Smith, was able to get to it and by then it had decayed and become so smelly that much of its guts had been taken out and thrown away.

Ever since then, Professor Smith had been searching for another. He had circulated leaflets in fishing villages all along the African coast offering a huge reward. But nothing. Until, Professor Huxley explained, last week. And then one had been landed in the Comoro Islands, a small archipelago few people had heard of, lying off the East African coast. Professor Smith had contacted the South African prime minister, who had given him the services of a plane from the South African Air Force. He had flown up to the Comoro Islands, collected the fish and flown back to Cape Town with it. The only trouble was that the Comores belonged to France, and the French were outraged at this violation of their territory – and by a war plane at that. As to the fish, well it still had its guts and the Professor would tell us all about them in due course. The fuss died down.

Twenty years later, I was still working in television. By this time, our programmes had become somewhat more professional and I was preparing a series that would tell the story of *Life on Earth*. What would we do about the coelacanth? By now a lot more specimens had been dragged up from the seas around the Comores. The fish certainly lived in deep waters, but not beyond reach. Nonetheless, no one had ever photographed a live one. Peter Scoones, a great expert in underwater photography was a member of the production team and he reckoned he could do so. He had one of the early small electronic cameras and could modify it for underwater use. And being electronic, it could send pictures that we could watch on a monitor on board ship. He had identified a patch of sea where the majority of the coelacanths had been caught. They were, he was sure, bottom-living creatures that swam up from the depths every night to hunt over this particular patch of the sea floor. It was far too deep for scuba divers, but if we could lower his electronic camera on a long line – a couple of hundred metres or so – down to the sea floor, there was a sporting chance that we could record pictures of one. And what a triumph that would be for *Life on Earth*! We decided to take the gamble.

We turned up in the Comores and recruited the most successful of all the coelacanth fishermen as our guide. And on our second night there, off we went to sea. The water where we were working was so deep that we could not anchor. But unluckily – indeed disastrously – there was a strong and steady current. Peter's camera worked all right. It gave us excellent pictures of the sea floor, which was covered with rocky reefs. But we could see no fish of any kind as we and the camera were carried along at a steady unrelenting two or three miles an hour. I almost wished that we would *not* see a coelacanth because if we did, all we would get would be a brief glimpse as we swept on by. But we kept at it. And then, on the third night, the inevitable happened. Although we did our best to heave up the camera as its pictures showed us it was advancing on a reef, it eventually hit one and got stuck. The cable broke. We had to abandon our one and only electronic underwater camera on the bottom of the sea. And that was that.

We packed up to go home. We had been filming other things during the day on the island itself – fruit bats – and Peter thought that there was still a little more he could do, so he decided to stay on for a couple of days. The day after we left, he was woken up in his room by the hotel porter hammering on his door. A fisherman had just arrived in the harbour with a coelacanth in his canoe and it was still alive. Peter grabbed his film camera, rushed down to the harbour and persuaded the fisherman to put his coelacanth back in the water. It was not long for this life. Groggily it drifted along the bottom with feeble movements of its fleshy fins while Peter swam all around it with his cameras. So the first pictures ever taken of a living coelacanth appeared in our film. In a way it was a triumph – but a limited one.

Twenty years later, a German film team took up the challenge. They decided to use a mini-submarine and from it they got the first really good shots of a coelacanth behaving naturally. It waved its fleshy-based fins in a lazy languorous manner, moving each independently. And then, as the mini-submarine came closer – and to the Germans' great surprise – it slowly tilted forward and stood on its head. And that happened not once but repeatedly. Why, no one is still quite sure. The one thing it didn't seem to do with the four fins on its underside was to push itself around over the sea floor in a way that some of its ancestors might have used in moving around on land.

But by now, hundreds of coelacanth specimens had been caught, dissected and studied and scientists had decided that although the fish were undoubtedly very ancient, they weren't after all the direct ancestors of the first land-living vertebrates.

And now there has been another discovery. Fossils of an ancient fish have been collected from 375 million year old rocks in the Canadian Arctic. It looks not unlike the coelacanth but belongs to a sister group. It is called *Tiktaalik* – after the local name for a huge freshwater fish that is found in the shallows in these parts. It too had fleshy fins strengthened with bones. But it also had a more mobile neck than any fish alive today and that no doubt would have been very useful in snapping up insects on land. Today, *that* is considered to be a more likely ancestor for us all than the coelacanth. So now I'm waiting for the news that a living *Tiktaalik* has been fished up from the deep sea somewhere or other. I can't wait to have another go.

In 1939, Marjorie Courtenay-Latimer was Curator of the small Museum in East London. In late December she saw down at the docks a very unusual fish that had just been landed. This is the rough sketch she made of it.

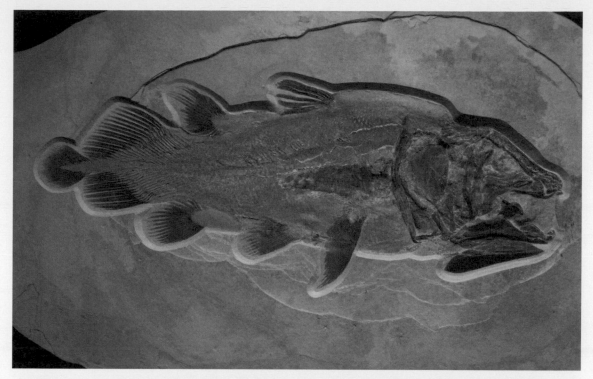

Coelacanths, like this fossilised one from Solnhofen, flourished during Jurassic times but the family was believed to have totally died out some seventy million years ago. The shape and number of the fins with their fleshy bases are highly characteristic.

Professor Smith triumphantly shows the body of the coelacanth to Dr. D.F. Malan, the President of South Africa.

Above, Professor J.B.L. Smith, the crew of the South African aircraft that took him to the Comores, and the French Governor of the islands pose beside the coelacanth.

Right, the first photograph ever taken of a living coelacanth. The fish had been brought in by a local fisherman and was barely alive, but Peter Scoones persuaded the fisherman to put it back in the sea so that he could photograph and film it.

In 1988 a German team, led by Hans Fricke and using a mini-submarine, at last filmed and photographed coelacanths behaving in a normal way. Since then, another population of coelacanths has been discovered off the coast of Sulawesi in Indonesia.

14

Dodo

The dodo is dead. Everyone
knows that. But why?
Because, fundamentally, it lost
its powers of flight.
Its ancestors, probably some
kind of pigeon, arrived on the
island of Mauritius in the
Indian Ocean, way back in
evolutionary history.
They were probably blown
there by a storm.

◁ The seventeenth century Dutch artist, Roelandt Savery, made a speciality of
painting dodos. He produced many versions and many artists since have based
their own vision of the bird on his. This one, in Oxford University's Natural History
Museum, was produced by Roelandt's son, Johannes.

The dodo is dead. Everyone knows that. But why? Because, fundamentally, it lost its powers of flight. Its ancestors, probably some kind of pigeon, arrived on the island of Mauritius in the Indian Ocean, way back in evolutionary history. They were probably blown there by a storm.

And it was a great place for a bird. There were no rats or foxes or cats or any other animal that might attack it. The island was big – fifty miles or so across – and well vegetated. So why bother to fly once you get to such a place? Flying is, after all, a very energy-consuming business. Indeed, it may have been better for them *not* to fly, for doing so might risk being blown away again by another storm. So the ancestral dodos didn't – generation after generation. And their wings, being no longer used, over many generations, dwindled in size and ultimately became so small that they could no longer lift the birds into the air. And they found all they needed on the ground. Pigeons eat all kinds of things and Mauritius had plenty of food – fruit, leaves, shellfish on the sea shore. At any rate, the dodos grew very big – even bigger than turkeys and much the same shape. And fat!

And then, some time around the last years of the sixteenth century, men in sailing ships arrived – Dutch, Arab and Portuguese. The dodos, big, fat, and flightless, were also defenceless. The sailors, after weeks and months at sea, were hungry, desperate for fresh food, for fresh meat. And the dodos, not only flightless but guileless and inquisitive from all accounts, didn't even flee. So it was only too easy for the sailors to club them to death. And that, it seems, was why the men called the birds 'dodo'. Because in Portuguese that means stupid.

Humans, having discovered the island, also began to settle on it, and with them came rats and cats and pigs and monkeys. Some forty years after the island's discovery, the inoffensive flightless dodo really *was* dead. We can't be sure of the exact date that the last one died, but certainly by 1690 they had all gone. Few physical fragments of their bodies remain today to prove that they had once existed. The last stuffed specimen was held by the Ashmolean Museum in Oxford, but by the middle of the nineteenth century it had become so decayed and tattered that the museum authorities decided it should be burnt. And all that survives now are bits of its head and its right foot – rescued from the flames by one of the Museum's curators.

That history has been repeated many times. Birds fly to a remote island, colonise it, lose their powers of flight – and when the world catches up with them, become extinct. And it also happened close to home. The cliffs surrounding islands in the northern seas around both America and Europe are still thronged with seabirds – razorbills, fulmars, and guillemots. But once, on the more inaccessible islets around Iceland and Newfoundland, Greenland, Ireland and Scotland, on cliffs that were very difficult for man or animal to reach, there were some birds that were bigger than all of them, Great Auks. They too in isolation had lost the powers of flight and used their wings only in swimming. They stood two and a half feet high, larger versions of their cousins, the razorbills, which still retain the ability to fly, albeit not very powerfully.

On Funk Island off the coast of Newfoundland, there was an immense colony of these huge flightless seabirds – at least a hundred thousand pairs of them. There, every summer men went out to slaughter the birds in thousands for the sake of their flesh and the rich oil that they accumulated in their bodies just beneath the skin. And once again, the species had no defence. One by one the Great Auk colonies disappeared. Those around the Orkney Islands and St Kilda in the far north of Scotland survived longer than most but by the middle of the nineteenth century, they too had all virtually gone.

In 1840, three fishermen from St Kilda were stranded by bad weather on Stac-an-Armin, a tiny rocky islet just off the main island. There, on a ledge, to their astonishment they came face to face with a strange giant bird of a kind they had never seen before. It was a living Great Auk. They caught it by the throat, tied up its legs with a rope and took it back to the little stone bothy where they were sheltering. The bird struggled so powerfully that it almost succeeded in cutting the rope with its bill. That night the weather worsened and a great storm blew up. As the wind lashed the little shelter, the bird began to shriek and the men became convinced that it was a witch that had brought the storm. Terrified they tried to kill it by beating it with two large stones. It was an hour before the bird died. They threw the battered body behind the shelter. And the storm subsided. That was probably the last Great Auk in Britain. Elsewhere, in remote islands, the bird managed to survive – but not for much longer. Another was killed on the tiny Icelandic island of Eldey between the second and the fifth of June 1844. And by 1855 the species was certainly extinct everywhere.

The name Great Auk, it seems, is a comparatively recent one. In earlier centuries people called the birds – believe it or not – penguins. The origin of that name is debatable but one explanation is that it comes from two Welsh words – 'pen' meaning 'head' and 'gwyn' meaning white, for the Great Auk did indeed have a large white patch on both sides of its head just in front of its beak. That, certainly was the name for it that was used by British seafarers during the sixteenth and seventeenth centuries. During those years and later in the eighteenth century, some of those men began to sail farther and farther south, beyond the southern-most tips of Africa and South America and farther still to the ice-capped islands of the Antarctic such as South Georgia. And in all those places, they saw immense flocks of large flightless birds with black coats, white chests and wings reduced to flippers that they used when swimming. What other name would they give them except the one they used for a similar bird they knew well up north. They called them, of course, penguins. For some reason, we no longer use the name for those northern birds even though they had it first. But it has certainly stuck to those in the south.

Antarctica is, of course, an island continent. Way back in geological time, when the dinosaurs dominated the lands of the earth, it together with South America and Africa, was part of a great super-continent. Eventually that broke up and the fragments we know today as continents drifted apart. Antarctica slowly moved towards the southern end of the globe and became so cold that an ice-cap developed over virtually all of it. But its coast and the ice shelves that extended from them across the surface of the sea, offered a valuable place where seabirds could land in safety when the time of the year came for them to lay their eggs.

Up in the north, the lands of Europe, Asia and North America extend from the warm south where it never freezes even in the depth of winter, up beyond the Arctic Circle. So every summer, when birds breed, land-living predators – foxes, wolves, bears – can travel north and raid the breeding colonies. There the ability to fly remains essential for birds.

But that was not so in the Antarctic, an immense continent isolated from all others by surrounding seas that are the roughest on the globe. Here was an island, but an immensely large one. And once again the evolutionary story repeated itself. Sea birds that once flew, now had land on which to build their nests and lay their eggs – land that was safe, free of land predators. So why fly? And the ancestral penguins which were probably closely related to albatross-like birds, also lost their ability to do so.

And then, a mere two hundred years ago, human beings caught up with them as well. We are now going to the Antarctic more and more frequently and in greater and greater numbers. The penguin colonies there are some of the most spectacular sights on the natural world. Will this story of flightless birds end in the same way? Perhaps – this time – we might take a lesson from history.

Sir Thomas Herbert travelled to Persia in 1627 accompanying the newly appointed British ambassador. On his way back he called at Mauritius and produced this sketch of the dodo. 'Her visage', he wrote, 'darts forth melancholy, as sensible of Nature's injurie in framing so great a body to be guided with complementall wings, so small and impotent that they serve only to prove her Bird.'

Funk Island in Newfoundland was one of the Great Auk's main breeding colonies. At its height it was estimated that it held 100,000 breeding pairs. This imaginary view of the colony was produced by J.G. Keulemans, a Dutchman by birth who settled in Britain in 1869 at the age of twenty-seven and became the most prolific bird-illustrator of his time.

By 1800 the Great Auks had been exterminated on Funk Island, so the bird hunters had to visit smaller less productive islands in order to continue the slaughter.

Stac-an-Armin, the rocky islet off St. Kilda. This was where the last Great Auk in Britain was killed.

Life in St Kilda during the last century when this photograph was taken must have been bleak indeed. Each year men collected nestlings from the seabird colonies for food. The last people to live there left in 1930.

15

Tracks

It is said that an Aboriginal
man, travelling with his family
group across the Australian
desert on their way to attend
an important tribal ceremony,
recognised a footprint in the
sand ahead of him as being
made by a cousin who he had
not seen for decades – that
print to him was just as
recognisable as a face.

◁ Desert living people, all over the world, must of necessity be expert trackers able
to detect and understand the significance of slight disturbances in the sand's
surface that are invisible to people reared elsewhere. None are more skilled at doing
so than the San people of South Africa

It is said that an Aboriginal man, travelling with his family group across the Australian desert on their way to attend an important tribal ceremony, recognised a footprint in the sand ahead of him as being made by a cousin who he had not seen for decades – that print to him was just as recognisable as a face.

How true that is I don't know, but I do know for sure that some human beings are able to interpret tracks with an accuracy and a wealth of detail that most of us would regard as uncanny – if not unbelievable.

It doesn't take superhuman powers to recognise that some particular tracks were made by a particular kind of animal, but marks on the ground can be much more informative than that. I once accompanied a group of Bushmen in the Kalahari Desert who were hunting kudu, the big antelope that is found there. They were soon on the trail of one. Where it crossed smooth windblown sand, the footprints were clear – even to my eye. But then they came to a stony gravelly patch of ground, and as far as I could see, they totally disappeared. The Bushmen however were not in the least baffled. To them, the prints were still clear and obvious. Not only that, they could see where the animal paused, what speed it was moving at, and how long it was since it had passed that way. All that information they gathered from clues that were totally invisible to my eye – the precise contour and dimension of the hoof print, its varying depth which indicated the angle at which the hoof met the ground, and how much dust had been blown into the print since it had been made.

It is the same in a forest but you have to learn at least the rudiments of that skill if you are not to get separated from your guide – and lost. You must be able to recognise the leaf that is out of place, the stem that is broken but has not yet started to exude sap and therefore must have been snapped only a few minutes earlier. I'm not much good at it and recall only too vividly following a man in the Amazonian rain forest who was going to show me the nest of a particularly interesting hummingbird. He moved through the forest not only silently but at speed and I was having trouble in keeping up with him, what with registering disturbed leaves and stems oozing sap, as well as dodging lianas. Wading across streams was a particularly tricky moment because then there is a gap in the track and you have to find the point on the opposite bank where you will be able to pick them up again.

I did my best to keep up. And then, in my search for clues, I noticed a rather odd liana. Surely I had seen one almost exactly like it only a few minutes earlier. I hurried on – and there it was again! When I saw it for a third time, I realised that the worst had happened. Those ahead were, understandably, not walking in a straight line. Every now and then they had to turn aside to avoid a particularly boggy area or a tangle of lianas that seemed to be more impenetrable than most. So the track had been zig-zagging and somehow or other I had walked in a circle and was now following my own track. So I had the humiliation of having to shout for help.

Interpreting the tracks of living animals is one thing. Doing the same thing for animals long dead is something else entirely. And there are wonderful tracks to be found in the rocks. Sometimes – and very exceptionally – you can be quite sure what made them. One of the most poignant ones I know was discovered in the famous fossil rich limestones of Solnhofen in Bavaria. One large slab is marked with two parallel lines of scratches about six inches apart. Between them runs a single intermittent groove. Follow them with your eye and you can see that they proceed along the lines of a giant question mark – straight at first then curving round to the left, coming round again until ultimately, at the end, you see the perfectly preserved shell of the animal that made them. It's a horseshoe crab. The long intermittent groove was made by the stiff spike of its tail, and the scratches on either side of it, the marks of its jointed crab-like feet. As it nears the end of its trail, you can see that it lifts up its tail spike and makes a series of repeated bar-like marks in the creamy silt. The animal is plainly distressed and – within a few inches – it dies. A single death, a hundred and fifty million years ago, graphically recorded for eternity.

That horseshoe crab trail is the exception in that there is no doubt whatever about the identity of the creature that made it. But that is seldom the case, when it comes to fossil tracks. Sometimes a track can be very mysterious indeed. In that same Bavarian limestone, there are long wide grooves with transverse ridges across them that look like miniature motorcycle tracks. They baffled people for a long time, before it was realised that they were made by the shell of ammonites that had been picked up by a sudden tidal current and rolled across the surface of the silty lagoon floor and sometimes even lifted so that they skipped.

The most impressive tracks, however, are those that were made by dinosaurs. Some, like the huge three-toed prints left by iguanodon, are quite abundant. Other deep circular pits running in long processions across sandstone were clearly made by a group of those huge hefty plant-eating dinosaurs like brontosaurs. In one place I know in Texas, they have been exposed by the eroding waters of a stream and lie a few feet below the rippling surface and they look as though they might have been pressed in to the mud only a few minutes earlier.

Here in Texas, such print trails are so abundant that at one time – and indeed perhaps even still – you could buy yards of them from fossil dealers. There is a story that a wealthy American from the east coast built himself a magnificent new house somewhere on the wide open Texas ranges. His landscape architect told him that he could buy twenty or thirty yards of tracks which he could reassemble in his new house's extensive gardens among the ornamental cactus – and in due course, that was done. When everything was finished, the proud house-owner invited his nearest wealthy neighbour around to show off his newly completed property. His guest saw these wonderful tracks. 'What in the world are those?' he asked. 'Well,' said the proud house-owner, 'those are dinosaur tracks.' 'Oh my God,' said his neighbour, 'I had no idea they would come so close to the house.'

But the most extraordinary, informative – and indeed moving – footprints must surely be those discovered by a team of palaeontologists back in 1978. They were excavating an ancient track-way near Laetoli in Tanzania. The tracks were made in a volcanic ash that had come from the eruption of Sadiman, a nearby volcano – an eruption that could be accurately and confidently dated from the radioactivity of the ash itself to around 3. 6 million years ago. There are many ancient tracks here. Large circular depressions with rims of mud curled upwards and outwards that may have been made by a primitive rhinoceros. The twin slots made by the delicate hooves of antelope. And there among them, a long trail of prints that are undeniably human. Fossil bones of early man dating from that exact period have been found, but the skeletons are fragmentary and the subject of much debate. Were the ancient hominids in this far distant period walking like chimpanzees, supporting themselves with the knuckles of their hands resting on the ground? Or were they standing fully erect. And there in the prints was the answer. Unambiguous and undeniable. There were two of these individuals, one slightly smaller than the other. A male and a female? A man and a women? They are moving so close to one another that they might well have been in physical contact. Were they walking hand in hand? And beside them, an even smaller one – a juvenile? Or perhaps I may legitimately say – a child. At one point the little group seems to pause. Perhaps they are looking back at the volcano erupting behind them that has blanketed the surrounding plains with volcanic ash across which they are walking. They pause and look at it. That much even I can deduce. And then they move on – and vanish.

As settlements spread across the United States during the nineteenth century, lines of huge mysterious footprints were discovered in the rocks and linked to the dinosaurs that had only recently been identified and named in Britain. These were unearthed in a small town outside Boston, Massachusetts in 1858.

A huge sauropod dinosaur walking across a newly exposed mud flat some 145 million years ago left a trail of its footprints. As they dried out, they hardened. Soon after the incoming tide covered them with mud. Over millions of years, the mud became rock and great continental movements buckled the horizontal strata to form the Andes. Now erosion has exposed the dinosaur's footprints once again.

The water in the Solnhofen lagoon was extremely salty and very warm. Animals from the open sea were occasionally carried in over the reefs by exceptionally high tides but they probably did not survive for long. Horseshoe crabs were among them. This one, judging from its tracks, was very distressed and repeatedly beat its stick-like tail creating a series of parallel lines in the silt before it finally died.

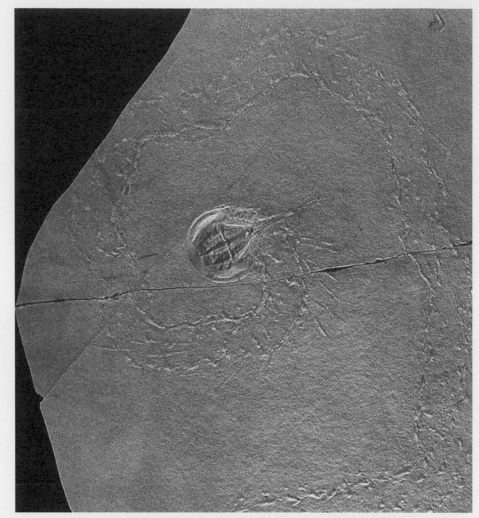

These mysterious marks are not uncommon in the Solnhofen limestones. It took some time before palaeontologists recognised what they are. Occasionally high tides in the Jurassic sea swept into the lagoon of Solnhofen. It picked up the ridged shells of ammonites lying on the muddy floor and bowled them along, creating these tracks.

166 Life Stories

In 1978 scientists excavating layers of volcanic ash that were deposited on the plains of Tanzania some five million years ago found fossilised hoof prints of antelope, primitive horses and what appears to be a kind of rhinoceros. Among them, there were these of human beings. Those on the right are significantly larger than those on the left. It seems likely that they were made by a male and a female walking alongside one another. And since there are no prints of knuckles, they must have been walking upright.

16

Birds' Nest Soup

Birds make their nests from
all kinds of materials. The
most common are twigs,
leaves, and mud.
Hummingbirds use spiders'
webs, cormorants – sea-weed.
And long-tailed tits construct a
lovely hollow ball.

◁ Whitenest Swiftlets nest on rock walls far from the entrances of Borneo's caves
where there is total darkness. The nest are often constructed within a few inches of
one another, proof in itself of how accurate the birds' aerial navigation must be.

Birds make their nests from all kinds of materials. The most common are twigs, leaves, and mud. Hummingbirds use spiders' webs, cormorants – sea-weed. And long-tailed tits construct a lovely hollow ball, and decorate the whole of the outer surface with fragments of a particular lichen. But one of the oddest and without any question one of the most valuable nests, almost worth its weight in gold, is that made by the swiftlets of southeast Asia. They use spittle.

There are number of different species of them. Some mix feathers and bits of vegetation with their spittle. But the most valuable nests are those made by the White Nest Swiftlets, because they make them out of spittle – and nothing else.

The nest is a shallow half-cup about four inches long. The bird, which as its name suggests, looks like a very small swift, black with paler undersides, flies into the farthest recesses of caves in Borneo where there is no light whatever and alights on the vertical rock walls, clinging on with delicate hook-like claws on its tiny legs – legs that are so short that they cannot even raise the bird above the ground if it is so unlucky as to land there.

Having selected a place for the nest, the bird produces the spittle from glands beneath its tongue and sticks a blob of it on to the rocks. Then, over the next few days, it returns repeatedly and adds more spittle, moving its head from side to side and ejecting it in long strands so that they stick together like a dense lattice to form the half-cup nest.

You may well wonder how on earth the bird manages to do all this in total darkness. Well, they find their way about by using a relatively simple form of echo-location. They make a clicking rattle and from the echoes, amazingly, they manage to find their way around the cave. During the day, when they are flying over the forest canopy collecting insects, their calls are high-pitched trills and they use their eyes, like any other bird. It is only when they enter the cave, as you can hear for yourself, that they start their rattle calls. These become ever more frequent as they go deeper into the cave and when finally, in total darkness they approach their nest, their rattling is almost continuous for then they have to navigate with an accuracy of a fraction of an inch if they are to land on their own nest and not on a neighbour's.

Once the nest is built, the female lays two small white eggs in it. One of the pair will incubate them while the other clings to the rock alongside. Filming this means getting up beside them and in the cave where we were working there was one fairly obvious way to do so. At the far end of the main chamber, a hole in the rock ceiling allowed a shaft of light to slant down and illuminate a strange grey dune that rose from the rock floor to within a few feet of the ceiling. Most mysteriously, its surface glinted and shimmered as though it had a covering of gold dust. Close up, it turned out that this dune was an immense pile of droppings – guano, as it is politely called – and the glints came from a moving carpet of cockroaches, their shiny wing covers catching the spotlight of sunshine.

The guano was produced by an immense colony of bats that covered the ceiling. As it accumulated on the floor, so the cockroaches munched their way through it to extract any remaining nutriment that it might contain. Among the cockroaches, here and there, were black sexton beetles. They were eating the bodies of the cockroaches when they died. And around the margin of this dune, where it met the rock walls, spiders had spread their webs to trap any of the cockroaches that might stray from the dune's surface.

To a biologist, it was a nice example of a direct and uncomplicated food chain which extracts every possible calorie from a single food source, the bodies of insects, tons of which were brought in every day in the stomachs of bats. The director in charge of the film we were making, explained that from the top of dune, it would be possible to get excellent views of the swiftlets that were sitting on the nests, stuck to the nearby walls. He added that it would be better, so as not to disturb the swiftlets too much, if Maurice, the cameraman, Hugh, his assistant and I went up to the top while he directed us from the bottom.

Obediently, Maurice, Hugh and I started to trudge upwards with our trousers sensibly tucked inside our socks. To start with, it wasn't too bad. The guano was crumbly, rather like soft, slightly sticky sand, but as we neared the top the ammonia rising from it became somewhat choking. 'Say something!' yelled the director from the bottom. I cleared my throat. Hugh turned on our battery light – and immediately the bats just above us unpeeled from the ceiling and started to fly around us in a huge rustling circling cloud. I looked at the camera. 'Many people,' I said, 'might be afraid that these bats would get tangled in one's hair. But the bats have an amazing navigational system based on echo-location and there is no chance whatever of them doing that'. I managed to raise a wan smile. Hugh turned off the battery light. And a bat hit me four-square in the face.

The navigational systems used by bats and birds are, however, a spectacular example of how two fundamentally different kinds of animal manage to solve similar problems. The first was how to fly. The birds did that, some two hundred and twenty five million years ago, by elongating just one finger in each hand and adding feathers that turned their arms into wings. The bats, over a hundred and fifty million years later, tackled the problem in a slightly different way. They spread out *all* their fingers, with skin stretched between them.

But then came the problem of aerial navigation in the dark. The bats solved that early in their history with echo-location and over tens of millions of years developed it into a highly sophisticated and marvellously accurate system using high-frequency signals beyond human hearing. Much later – and probably very recently indeed, in evolutionary terms – the swiftlets, faced with the problem of flying in the dark when they began nesting in caves, started to develop a navigational system based on the same principle of echo-location. But since their system evolved comparatively recently, it is not surprising that it is simpler than that of the bats.

We managed to get an even better view of the swiftlets with the help of a group of men from the nearby village who were working in the cave. They came here every year to gather the swiftlets' nests by climbing up to heights of a hundred feet or so, using bamboo poles and rattan ropes, and knocking off the nests with sticks. It seemed to me appallingly dangerous and indeed people do get killed every year in collecting this bizarre harvest. But men continue to do so for the nests are sold to visiting traders for extraordinarily high prices.

They are used to make that favourite of the Chinese gourmet, birds' nest soup. The trade is an ancient one. Recipes for making the soup date from the 7th century. It was not only thought to be delicious but good for you. It was tonic and therapeutic. It cleared phlegm from the throat and soothed gastric problems. But it seems that Chinese gourmets also found – and still find – an intellectual pleasure in eating food that is bizarre and exotic. And what could be more bizarre and more exotic than the nest of a bird that haunts the caves on the other side of the South China Sea in far distant Borneo. So birds' nest soup became an essential course in the Emperor's banquets. And the taste for it has continued to this day.

Before the nests are cooked, they have to be cleaned. This is done by soaking them so that any foreign bits and pieces such as feathers can be picked out from them. When that is finished, they are a brownish white and slightly translucent. Then they are mixed with chicken and pork and that, together with spices and water, creates a slightly slimy soup.

When we at last returned from the caves to the nearby town, I thought that I really ought to discover what a bird's nest tastes like. That was not easy. The restaurant owner couldn't understand why I should want to taste the nest by itself but eventually he produced one for me. Well, it may indeed be a great boost to one's health, provide general therapeutic improvement and even, as the chef assured me, do wonders for my sex drive. But I can tell you, by itself a cave swiftlet's nest, properly cleaned, – to my palate at any rate – tastes of nothing whatsoever.

This is the nest of a Blacknest Swiftlet. These birds nest nearer the cave entrances where there is a little light to help in navigation. Their nests, however, are not as highly valued as those of the Whitenest species, for they contain a lot of their makers' feathers, all of which have to be removed before the nest is ready for the kitchen.

The cave system of Gomantong in Sabah, Borneo, is populated by both Blacknest and Whitenest swiftlets. Each year some five tons of the tiny nests are taken from them by men working from a rickety network of rope ladders.

△ Immense colonies of leaf-nosed and horseshoe bats cover great areas of the ceilings of Gomantong caves. It is they that produce the steady rain of droppings that accumulates on the cave floor in a huge dune. At one point its summit almost reaches the ceiling. At its base, there is a wide hole in the rock floor through which the guano is steadily flowing, presumably into another lower chamber. Understandably, no one has yet chosen to enter it.

Good though the bat's digestion may be, their droppings still contain some nutritive value and the surface of the dune is covered by cockroaches munching their way through it. With them are black beetles that consume the bodies of the cockroaches when they die. And there is still enough nutriment in the guano for it to be sold for high prices as a superb agricultural fertiliser. ▷

A pair of Blacknest Swiftlets at the nest. Male and female are identical so it is not known whether incubation is undertaken by one or both of a pair. But certainly the two spend the nights together at the nest.

17

Amber

I had a small museum as a boy – as many boys do. Mine contained a Victorian bun penny, the shed skin of a grass snake, some fossil ammonites, and a twisted piece of metal from an incendiary bomb that had been dropped on Leicester.

◁ The rarest of all amber inclusions, a backboned animal. Most creatures of any size in the amber forests would have had the strength to free themselves from tree resin. This little gecko, preserved in Dominican amber, failed. Perhaps it was already dead having been caught, killed – and perhaps dropped - by a predatory bird.

I had a small museum as a boy – as many boys do. Mine contained a Victorian bun penny, the shed skin of a grass snake, some fossil ammonites, and a twisted piece of metal from an incendiary bomb that had been dropped on Leicester where I lived – for this was during the war – and a small brown object, two and a quarter inches long.

I can be quite precise about its measurements because I still have it. It is roughly rectangular with curved ribbed contours that have been beautifully polished. It is not cold, like stone, but warm to the touch. Lift it up – and I can see through it. And there within, at one corner, lies a small fly. It is, of course, a piece of amber and I was given it by a girl, a refugee from Nazi oppression, who in 1938 stayed in our house for a few weeks on her way to relatives in the United States. She had come from a town not far from Gdansk on the Baltic Sea.

Gdansk is a city that is known all over Europe for its amber. There is a street there lined with shops – some big, some no more than wayside stalls – that sell nothing but amber. People living on this coast have collected and traded in amber since the Bronze Age six thousand years ago. They find it on the beaches, cast up by the waves. Being lighter than sand and scarcely heavier than water the pieces are deposited on the wet surface of the sand as if they are gifts from the sea.

The Greeks knew amber well and were puzzled by it. It was soft enough to be scratched by metal. So it was not a gem, even though it was as beautiful as many a precious stone. And they noticed that it has an extraordinary property. Rub it with a cloth and small pieces of paper or chaff are so attracted to it that they will lift off the ground or even fly through the air to attach themselves to it. The

Greeks called this wonderful substance 'electron', a word that three thousand years later was borrowed by western science in describing the force that enables amber to lift paper and which mankind eventually harnessed and called electricity.

But what about the insects such as the fly in my piece? Look at it with a lens, and you are in no doubt that this is a real fly, albeit quite a small one. It is complete in every detail. Such inclusions take some explaining. But those sharp-minded pragmatic Greeks correctly reasoned that at some time, this extraordinary substance must have been soft and sticky and deduced that it was tree gum that somehow had turned to stone.

Resin, like that oozing from a cut in the trunk of a coniferous tree today, contains quite a lot of aromatic oils which evaporate – and in doing so produce that delicious smell. As that goes on, so the gum hardens. And it continues, very much more slowly, even when that gum has solidified into a lump of amber. As a result its polished surface will eventually become slightly crazed with a network of tiny cracks and – after centuries – develop a powdery rind. Ladies who wear amber necklaces know that if it is to retain its surface, it needs the gentle polishing that comes from regular wear – or at least a regular rubbing with a cloth.

It is not just flies that you can find in amber but tiny scorpions, seeds, mosquitoes, spiders, mites, bits of leaves, ants – lots of ants – and in a few exceptional and extremely valuable pieces, creatures as big as small lizards. If you do have a piece with a lizard in it, then you should look at it very hard. Fakers have been known to take a piece of amber, excavate a little hole in it and then insert a suitably mangled fragment of lizard – its front half, perhaps. Then they fill the space around it with a synthetic resin that has exactly the same optical properties as amber. You will have to be very sharp-eyed indeed to spot the deception.

Genuine amber comes from mudstones that were deposited some forty million years ago and are now outcropping on the bottom of the shallow Baltic sea some miles out from the coast. As currents erode the seafloor – and particularly during a storm – the mudstones are turned to silt and the lumps of amber, being lighter than the mud, are lifted up, carried across the sea floor and finally washed up on the wet glistening sands of the beaches.

For centuries, Europeans thought that the Baltic was the only place that produced amber in any quantity. But there are sources of amber in the New World too. Many of the indigenous people – the Maya, the Aztec – also knew and treasured the material. The main source there today is up in the hills of the Dominican Republic, a state that occupies the eastern half of the Caribbean island of Hispaniola.

Dominican amber is somewhat younger than that from the Baltic – around twenty six million years old. The creatures it contains are generally much the same – ants, flies, scorpions and so on – though an amber expert could spot the differences, just as an entomologist would be able to tell the difference between a European mosquito and an American one.

A few years ago, I went to the Dominican Republic to have a look at the way the amber is collected there. Up in the misty hills, half naked men dig narrow and, I am sure, extremely dangerous shafts into the mountain side, hacking away at the blue mudstones by the light of oil lamps. Because the lumps of resin weighed less than the mud of the swamps in which the trees grew and ultimately fell, the tides swilling through the swamps tended to pick up the resin lumps and drop them together in slack water away from the main current. So now the amber often occurs in seams and once a miner finds a piece, he will carefully trace the seam in which it lay, knowing that that level is likely to produce more. The miners themselves roughly polish the pieces on simple treadle-driven lathes so that they can get some idea of the quality of the amber they have found and know whether or not it contains anything particularly valuable. Every evening, traders come up from the little town on the coast to buy it.

We filmed there for long enough to get to know who was who and one evening, when we were sitting having a drink in the main square of the little town at the foot of the hills, I noticed one of the main traders arrive, carrying a cloth bag which he put on the table in front of him while he ordered a beer.

I went over to have a few words. Had he found anything particularly interesting? He didn't know. He hadn't properly sorted through the day's take. Eventually, after a lot of haggling, I took a gamble and bought the whole bag. So now I have several hundred pieces of Dominican amber to put alongside the Baltic pieces I have acquired over the years. Of course, I have had a look at all of them with a hand lens. But to investigate a piece properly takes time.

As I sit late at night looking down a microscope at a piece I have not investigated before with proper intensity, I can imagine myself as a naturalist, wandering through a forest that flourished thirty million years ago. Of course, I cannot hope to find any animal bigger than a fraction of an inch. Nothing of any size is likely to get stuck in tree resin. Sabre-toothed tigers and mammoths, which were certainly around at that time, didn't climb trees. But experts have found mammalian hairs and identified them as coming from squirrels and dormice and marsupial shrews.

Mosquitoes, of course, were abundant and they are common in amber. Now when mosquitoes bite you, me or anything else, they suck blood and fill their tiny stomachs with it. Twenty odd years ago, there was a suggestion that we might find dinosaur blood in a mosquito's stomach, extract genetic material DNA from it and then bring back to life, if not a whole dinosaur, then perhaps a bit of dinosaur tissue.

In fact, dinosaurs had disappeared long before the time that Baltic or Dominican amber was forming. And alas! such genetic material as has been extracted from amber has been too badly mangled to offer any hope of resuscitation in any way. To be truthful, I don't mind that much. Examining a swarm of ants, their little bodies gleaming gold, and complete with the tiny hairs on their heads and minute glinting facets in their eyes, is wonder enough.

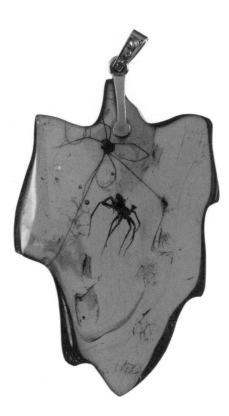

Jewellers, instead of rejecting pieces of amber with inclusions as being imperfect, sometimes take a delight in the beauty of the creatures they may contain. This charming Victorian pendant contains a spider and above it, a daddy-long-legs.

Right. The trees that produced Baltic amber 30 million years ago in the Miocene period are thought to have belonged to the pine family. They would have produced resin like this, flowing from the damaged trunk of a conifer. Dominican amber on the other hand, probably came from an extinct leguminous tree related to a living Caribbean species known locally as algorroba.

Left. In the Dominican Republic miners dig deep shafts in the hillside, and search for amber in sauna-like conditions of high temperatures and humidity.

Right. Amber has been treasured by humanity since prehistoric times. This elegant cup, 3½ inches across, was discovered in a Bronze Age grave near Hove in Sussex. Its date has been estimated as between 1285-1193BC and it is thought to have been the treasured possession of a particularly powerful chief and to have been buried with him.

Amber 185

Many creatures were trapped in resin while they were in the middle of their own dramas. These midges, each enclosed beside a bubble of air, were in the middle of mating, with their abdomens attached.

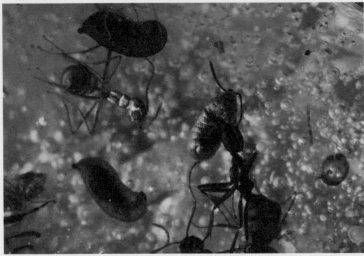

These ants must have been in the process of moving their nest, for one of them is carrying a grub in its jaws.

Ants tend to follow one another's trail but in this case it led to resin. The leader has got stuck. The follower has grasped it by the abdomen and appears to be attempting to pull it free, only to have become stuck itself.

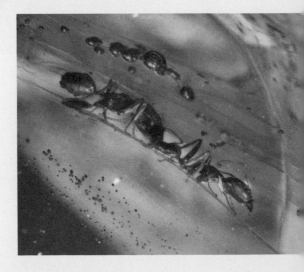

Amber preserves with such perfection that it is possible to examine the anatomy of some of the creatures it contains in microscopic detail. The shape of its antennae shows that this mosquito in Dominican amber is a female. Unlike males, females suck blood, usually from backboned animals such as reptiles, birds or mammals. This particular one has clearly been using her long thread-like proboscis to do that for her abdomen appears to be full. Theoretically, it might be possible to extract DNA from such blood. It could not, sadly, have come from a dinosaur for they had long been extinct, but it might nonetheless have come from some large dramatic animal such as a sabre-tooth tiger.

Amber 187

18

Large Blue

The Large Blue butterfly is
actually quite small, about an
inch and a half across.
It is, however, bigger – just –
than the Common Blue
or any of the other species of
blue butterfly that you might
see fluttering in your garden
or over the sun-baked turf of
the hills and heaths of
southern England.

◁ The Large Blue is comparatively small with a wingspan of about an inch and a
half but it is one of the most beautiful of British butterflies. For many years it was
also one of the most prized by British collectors – before that activity became illegal.
But in 1979, in spite of the rigorous protection of its few remaining colonies, it
became extinct.

The Large Blue butterfly is actually quite small, about an inch and a half across. It is, however, bigger – just – than the Common Blue or any of the other species of blue butterfly that you might see fluttering in your garden or over the sun-baked turf of the hills and heaths of southern England – the Silver-studded Blue, the Chalkhill Blue, the Adonis and the Holly Blue. During the nineteenth century it was also the rarest of the Blues. Those two qualities – the rarest and the biggest – are, of course, an almost lethal combination for any animal. Collectors find the coupling irresistible and during the nineteenth century, when butterfly collecting was a passion that possessed not only schoolboys but doctors, lawyers, clergymen – in fact almost every kind of Victorian gentleman who had time to spare – the Large Blue became one of the most sought after of all British butterflies.

The zeal with which these collectors pursued their prizes was extraordinary. One clergyman wrote in triumph to a fellow enthusiast saying that he had found a Large Blue colony and had taken a hundred specimens from it in one single day. Not surprisingly, the Large Blue became rarer and rarer and by the middle of the last century, it was clear that the species was approaching extinction in this country. Conservationists identified the last remaining colonies. Fences were put up to protect them. Cattle and sheep were rigorously excluded. Wardens kept watch to guard against illegal collectors. But the numbers of butterflies in these last colonies got smaller and smaller each year. It was as though no matter how carefully the Large Blue was protected, it had developed a death wish. And in 1979, the last known colony – on Dartmoor – disappeared. Britain had lost its Large Blue.

Just before that happened, however, a young scientist, Jeremy

Thomas, who was working for his doctorate, was given the task of studying the last survivors. For six years, to use his own words, he lived beside the Dartmoor colony, recording everything visible and measuring everything measurable. What he discovered was not only complicated but truly extraordinary.

The adult Large Blue lays its eggs on the plants of wild thyme. The caterpillar, when it hatches, burrows into the thyme's flower bud and feeds on the developing seeds. If there are several caterpillars in the same bud, as there may be, then they become cannibals, eating one another until only one remains. As this one grows, it moults until eventually, with a flick of its body, it falls to the ground and crawls into a crevice in the earth. Before long, almost inevitably, it is found by a worker ant. The ant becomes hugely excited, crawling all over the caterpillar, licking its skin and sipping liquid, nectar, that the caterpillar produces from a small gland at the end of its body. Other ants may arrive to join in the feeding frenzy, but as far as the first one is concerned it is finders-keepers and eventually it drives away all the others. Then the caterpillar, after having been milked for some time – on occasion as long as four hours – suddenly hunches itself up so that its otherwise somewhat flabby body becomes taut. Now it has much the same feel as an ant larva and what is more it is much the same size. The ant certainly seems to think that this is what it is, for it picks up the caterpillar and carries it back to the nest.

Once there, the caterpillar makes its way into the chamber containing the young ant larvae and starts to eat them. It crawls over its victim, bites its skin and sucks out the contents of the larva's body. It does this repeatedly in bursts of activity, and as the weeks pass, it grows fast. The ants may, at this stage, realise that they have a stranger in their midst – in which case they will kill it. But if the caterpillar survives for five weeks or so it will have become so big and developed such a tough skin that it can withstand any attacks.

ant pupae. But eventually when it pupates, what emerges from the chrysalis is not a butterfly, but an ichneumon.

A similar but slightly different species of ichneumon is known to have parasitised the Large Blue. Specimens of it have been found neatly pinned in the cabinets of those Victorian collectors. So the question arises – should we welcome such an ichneumon back so that it might resume its relationship with our own Large Blue? In my view, we certainly should. The beauty of our Large Blue is incontestable. But one of the most wonderful things about the natural world, for me, is its marvellous intricate complexity. And what could exemplify that more astonishingly than the intertwining of the lives of the *sabuleti* ants, the ichneumon and the Large Blue butterfly.

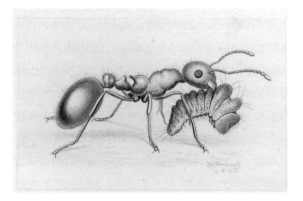

In the 1920's, the great British lepidopterist F.W. Frohawk spent over an hour with a lens in one hand and a pencil in the other in order to produce this remarkably accurate drawing of an ant transporting a Large Blue's caterpillar.

A female Large Blue may lay several small white eggs in the flower bud of wild thyme. But only one will survive, for at this stage, the Large Blue is a cannibal.

After the Large Blue caterpillar has moulted four times, it drops to the ground. At this stage it is about the same size as an ant grub. Furthermore, it produces a chemical smell that exactly matches the smell of a larval ant, and when a worker ant of the *Myrmeca* genus finds one, it takes it back to its colony.

Once it is within the ants' nest, the caterpillar starts to feed on the ant larvae, seemingly undetected by the worker ants, providing the nest it has reached belongs to the particular species whose recognition odour it mimics.

After some months in the ants' nest, the Large Blue caterpillar is many times bigger than the ants themselves. Nonetheless they treat it as if it were their queen. It will continue feeding on the ant larvae that lie helplessly around it for a year – and sometimes even two – before it turns into a chrysalis.

In continental Europe, the Alcon Blue caterpillar, a close relative of the Large Blue, is parasitised by a species of ichneumon wasp which injects the caterpillar with one of its eggs as it is doing here. As a result while the caterpillar feeds on the ants, it is being eaten from the inside by an ichneumon larva.

In midsummer, within the ants' nest, the skin of Large Blue chrysalis splits and from it emerges an adult butterfly – provided it hasn't been parasitised by an ichneumon.

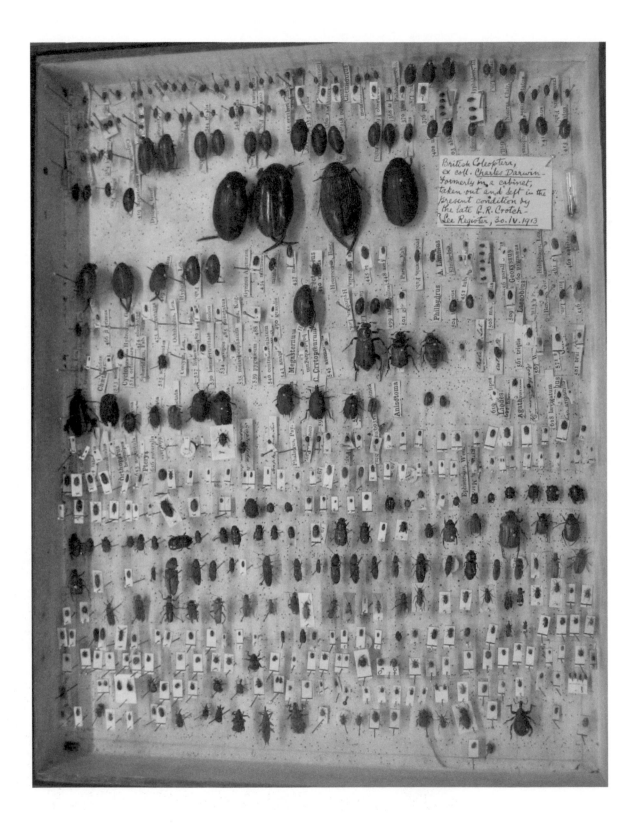

British Coleoptera,
ex coll. Charles Darwin.
Formerly in a cabinet,
taken out and left in the
present condition by
the late G. R. Crotch.
Lee Register, 30. IV. 1913

19
Collecting

Collecting is a strange
affliction. I have to admit to
being a sufferer since
childhood – stamps,
magazines that were
numbered in sequence, bus
tickets, coins, fossils –
and advancing years
have not really cured me.

Collecting is a strange affliction. I have to admit to being a sufferer since childhood – stamps, magazines that were numbered in sequence, bus tickets, coins, fossils – and advancing years have not really cured me.

So today I collect – among many other things – books about New Guinea and if I find one, no matter how boring it is, I am likely to buy it for no better reason than that I don't have it.

Where does this urge to collect come from? Some animals, certainly, collect objects but all those that I can think of collect things that have a use – caddis fly larvae collect tiny sticks with which to build the little tubes in which they live and will collect coloured beads, if you keep them in an aquarium and provide them with nothing else. Bowerbirds, the most spectacular collectors among birds, also assemble coloured objects but that is in order to create a display that will impress females. Human beings on the other hand, collect things that have no practical use and often don't even show them to anyone else but keep them secret, hidden away in a back room.

But why? Psychologists have produced a variety of answers – such as over-severe potty training in infancy, for instance – which I personally don't find very convincing. I have a different theory.

It seems to me that the affliction – if it can be called that – is by and large more masculine than feminine. There have been, it is true, one or two spectacular women collectors – Lady Charlotte Schreiber, for example who had a passion for little china figures of shepherds, shepherdesses and such like and left thousands of them to the Victoria and Albert Museum. But that is nothing compared with Sir Thomas Phillips who, in the 19th century collected books in quite phenomenal numbers. Most of us, I dare say, are guilty of buying more books than we actually read. But he bought forty or fifty a week and by the end of his life had a collection of forty thousand of them as well as sixty thousand manuscripts. And certainly most active collectors I know who scour shops, auctions and car-boot sales for the objects to which they are addicted are men – and men whose wives look at them with an affectionate even pitying tolerance when they spend yet more extravagant sums on something that is quite useless but which appeals to them irresistibly because they haven't got one exactly like it.

The male emphasis, I believe, is an important clue. There can be little doubt that there was a division of labour between the sexes early in mankind's history. The long period the human infant needs before it is capable of even walking by itself, let alone finding food, meant that women, by and large, remained in camp or cave, and men went off hunting for meat for the family. So the hunting instinct – the delight in finding prey, tracking it and catching it – is deep-seated in men. Indeed, it seems to me to be possible that men found a positive pleasure in the process and did not go off hunting only out of a sense of duty and responsibility towards their families. In short, I think the process of collecting objects is a way of satisfying the deep-seated urge to hunt – an urge that in modern life is not properly satisfied when all that a man brings back to support his family is a piece of paper or simply the information that a message has been sent to his bank.

Natural history objects – shells, birds' eggs, fossils, odd stones, skeletons – have been collected by people since the beginnings of scholarship. In the sixteenth century, Aldrovandus, the Italian author of the first great encyclopaedia of natural history, was said to have had four thousand five hundred and fifty drawers of specimens. Noblemen throughout Europe had their cabinets of curiosities in which they displayed anything – animal, vegetable or mineral – that seemed strange and remarkable to them. In the nineteenth century, Lord Walter Rothschild, fuelled by his family wealth, assembled the biggest collection of natural history objects ever made by one man, paying over four hundred collectors to scoop things up for him from all over the world. Giant tortoises, bird skins, birds' eggs, butterflies, beetles – there seemed to be no product of the natural world that he was unwilling to acquire.

Charles Darwin, in his youth, was a passionate, fanatical collector of beetles. As an undergraduate in Cambridge he searched for them obsessively. 'No pursuit gave me more pleasure', he said. He didn't dissect them. He simply classified them. That is to say, he learned to recognise different species. He arranged them, both in practice and in his mind, in some sort of order. He put those that were most like one another close to one another. He divided them into families. And that process must have made him wonder why there were so many species and what processes might have brought them into existence.

He was still at this stage when he was invited to join the *Beagle*, the naval surveying ship that was about to set off on a round-the-world voyage to survey the coasts of South America. But he did *not* go as a beetle collector or any other kind of naturalist. His official job was simply to be a companion to Captain Robert Fitzroy, the autocratic and irascible commander of the ship and to provide him with gentlemanly conversation.

But the collecting mania still possessed him. Everywhere the *Beagle* went, young Mr Darwin eagerly went ashore and collected – fossils, plants, mammal skins, shells, everything natural in fact, that was collectible. And it was that passion and those collections that gave him the raw material for the theory of evolution by natural selection.

It may come as a consolation to some of us that, on occasion, even the great Darwin was less than perfect as a scientific collector. It is said that the idea of natural selection was sparked in his mind by the claim made by a British resident in the Galapagos Islands that he could tell which island a giant tortoise came from by the shape of its shell. Those on drier islands which lacked a reasonable turf on which to graze had shells with peaks at the front of the shell that allowed their owners to crane their exceptionally long necks upwards and browse from the branches of tall plants. Darwin certainly brought back several shells and skeletons of these extraordinary reptiles but he had done the unforgivable. He had neglected to note which of them came from which island. So he couldn't use them to illustrate his theory. Instead he had to base it on the rather less dramatic mockingbirds that his assistant Symes Covington had not only collected but had meticulously labelled with their place of origin.

Darwin's son inherited his father's collecting mania. But in 1840 a new collecting possibility had arrived. Britain had invented the postage stamp and it had spread around the world. In 1862 Darwin wrote to one of his scientific correspondents, Asa Gray, the Professor of Botany at Harvard in the United States and asked him if he could possibly send his son some stamps. Not any old stamps, of course, but the Wells Fargo Company Pony Express tuppenny and fourpenny ones.

Stamps were still the rage when I was a boy but I sense that these days the passion has lessened with the sheer abundance of different issues. Very few contemporary ones are rare enough to quicken the collector's pulse. Modern marketing methods take care of that. Bus tickets, which back in my boyhood had different colours for different values, have now gone. Even train numbers, which were once in vogue, are no longer, I'm told, very interesting. More seriously, collecting many kinds of natural objects is now forbidden by law. For very good reasons, it is now illegal to collect birds' eggs or pluck rare wild flowers. Nor is it allowed, on many sites of geological importance, for a boy without a permit to go in search of fossils as I once did.

And I worry about that, for it seems to me that the collecting impulse was responsible for stimulating an interest in natural history and ultimately giving people a love and an understanding of the natural world. Maybe some of us will be able to translate that passion to accumulate material objects into an equally satisfying way of collecting photographic images of birds and butterflies, dragonflies and flowers. I hope so.

But there is no need for us to feel too guilty about the passion for collecting. For many of us it was the trigger that led us to the deep pleasures that come from an involvement with the natural world and an understanding of how it works. And it led one man of genius to propose the most important and revolutionary theory in the whole of natural science.

While a student at Cambridge University, Charles Darwin
became a fanatical collector of beetles.
One of his friends and fellow beetle-enthusiasts, Albert
Way, drew this cartoon of him pursuing his hobby.

Lord Walter Rothschild was one of the most energetic and voracious of collectors. As well as a vast collection of bird skins he had a gigantic collection of butterflies. Some of these drawers contain his favourites, the huge birdwing butterflies of the western Pacific and South-east Asia.

Lord Rothschild collected living animals as well as dead specimens. He had a herd of giant tortoises and also kept zebra and trained them to pull the carriage in which he travelled occasionally in London.

From the sixteenth century onwards, it became increasingly fashionable among European noblemen to own a cabinet of curiosities, in which they assembled all kinds of animals, plants, minerals, fossils and everything strange. One of the first was that belonging to Ferrante Imperato of Naples. This is the frontispiece to his catalogue showing him guiding visitors around his collection.

RITRATTO DEL MVSEO DI
FERRANTE IMPERATO

Above. A herd of giant Galapagos tortoises. Darwin failed to note the detail, critical for his theory, of which islands his specimens came from.

Below. The *Beagle*, having visited the Galapagos and crossed the Pacific with Darwin aboard, landed in Sydney, Australia. This sketch shows her when she returned to the same harbour five years later.

Charles Darwin aged 31, four years after his return from his voyage with HMS *Beagle*. He had been recruited for the voyage not as a naturalist but as a companion to Robert Fitzroy, the Captain. Nonetheless, he made observations and collections that were to revolutionise the natural sciences.

20
Adam's Face

Eyebrows are very odd.
What are they for? To prevent
sweat from dripping into our
eyes? Hardly likely. For one
thing they are muscled so that
we can raise and lower them.
We can even raise them
alternately.

◁ A Biami man from the mountains of central New Guinea. He had seen no
European face before. Nor did he speak any language known to any of the members
of the expedition who met him. Yet his concern and apprehension was clear from
the expression of his face.

Eyebrows are very odd. What are they for? To prevent sweat from dripping into our eyes? Hardly likely. For one thing they are muscled so that we can raise and lower them. We can even raise them alternately. And they are boldly coloured and often remain so, even when the hair on our head has turned white or disappeared altogether.

Simple sweat shields would hardly require such versatility or such permanence. An alternative explanation is that they are for communication. But if that is so, what do they say? One suggestion is that they can be used to send a message of acknowledgment. If you raise them quickly – and we all can – while looking at someone who is perhaps too far away to conveniently address with speech, then it is a sign that says you have registered the other person's presence and are content that they should be around. Of course, you can elaborate that message. You can stretch your mouth and turn it up at each end – that is to say, you can smile. And that adds the information that you are glad to see them.

So was the eyebrow-lift part of the ancient gestural language of mankind? Human beings must, after all, have had a system of communication before they elaborated their grunts and yells and other vocalisations into anything as complex as a true language. All other primates – monkeys and apes – use facial gestures and the human face is more elaborately muscled, more mobile and therefore more expressive than any of them. If that is so, then the eye-brow lift is not something we learn to do. It's instinctive.

When I was in Fiji some years ago, I thought I would test this notion. Here, after all, live people of many different races – Fijians, Indians from Asia, Africans and Europeans. So one morning I went down to the market and stationed myself beside the huge banyan tree that stood there, and as people walked towards me in the milling crowd, I caught their eye and performed my eyebrow lift, accompanied by a warm smile. It worked only too well and I quickly decided that I had better stop before I made too many and too many intimate friends – or was taken away by the police.

And then, some years later, I had an opportunity to test the idea rather more sensibly. Today, there are very few human communities who are totally cut off from the rest of humanity. But back in the 1970's there was a part of the world where they could still exist – the centre of New Guinea. The forest there is so thick, the mountains so rugged, and travel, consequently, so difficult, that no outsider had bothered to go there. The people living high up one of the rivers however said that there were nomadic people living in the forests farther into the mountains who they called the Biami. But they hardly ever saw them. Then a mining company applied for permission to prospect there and the Australian administration that was then in charge, decided that they had better send in a patrol to see if the Biami really existed. And I managed to get permission with a camera team to go with them.

It was exploration as it had been in the classic days. The terrain was so mountainous and the forest covering it so thick that there were no tracks of any kind, let alone any that could be traversed by any form of wheeled transport. Nor were there any clearings on which a helicopter could land. The only way to travel through it was that which had been used since the very beginning of exploration – that is to say, on foot.

Furthermore, the forest could provide very little food – if any. There were no large animals we could hunt for meat. Nor were there any fruits that we could be certain were edible. And we could not rely on getting any supplies from local people for we might not, after all, meet any. So we would have to take all our food with us.

Laurie Bragge, the local District Comissioner, who would lead the patrol, had a formula for working out how many porters we would need. Every man who was not carrying food – and that meant not only us but those men carrying tents, camping gear and our filming equipment – would need two others with him carrying nothing but food. The three of them would then have enough provisions to last a fortnight. Every day we stayed out longer than that increased the number of food carriers required so rapidly that it soon became impossible. Laurie calculated that since we expected to walk for at least a month, we would not only need a hundred porters, but halfway through the trip we would have to have an air drop of bags of rice.

It was as exhausting a trip as I have ever made. The vegetation was thick and tangled; the ground steep, muddy and slippery; everything you tried to hold on to seemed to be armed with spines; and it rained every day from midday on. We had interpreters with us – 'turnim-talks' as they are called in pidgin – but we needed three of them. Laurie spoke in pidgin to the porters. At the highest village up-river we had picked up a man who didn't speak pidgin but understood the language used by the porters among themselves. And he knew another language called Bisorio, which we hoped might be one that was understood by the Biami – if they existed.

After two weeks of walking through the empty forest, one of the carriers noticed human foot-prints. There were people here after all, and they were travelling just ahead of us. We followed the trail as fast as we could. That night, when we made camp, we put out gifts – glass beads, knives, cakes of salt – and one of the carriers sat beside them calling out 'Biami' every few minutes until nightfall. But in the morning the gifts were untouched. We followed the trail for the next four days, but the people ahead of us, whoever they were, were travelling faster than we could with all our stores and equipment. For the next four days we tracked them. Every evening, we put out gifts. And every morning they were untouched.

Then on the fifth morning, lying beneath the tarpaulin under which each of us slept, I opened my eyes and saw seven small men standing staring at me. They were naked except for a bark belt with sprigs of fresh leaves stuck into it, front and back and thin cane wound round their waists. Shells hung from their pierced ear-lobes and small twigs were stuck in punctures at the end of their noses. Here, if ever, was a moment for the eyebrow flash. I tried it. And the smallest of the men, who seemed to be their leader flashed his eyebrows back in response. But he didn't smile.

By now Laurie was up and had joined us. So did our turnim-talk who spoke to them in what I presumed was Bisorio. But it seemed that the strangers didn't understand him – or he, they. And yet we were able to communicate surprisingly well with gestures. Using a mixture of nods, smiles and pointing with our fingers, we managed to explain that we needed food and that we would give salt or beads in exchange. The beads did not interest them, but they licked fingers, tasted the salt and grinned delightedly.

Laurie started to question them about the number and names of the rivers that we had crossed. I was surprised that he should choose such a subject as the topic of conversation at such a time. But as their leader listed them, he counted them. One, he touched his wrist; two, his forearm; three, his elbow; and so on, all the way up his arm and shoulder until he got to eleven which he indicated by tapping his neck. The reason for Laurie's interest, I discovered, was not so much the rivers themselves as the way the people counted them. Counting gestures differ among New Guinea people and Laurie was trying to discover to which of the mountain people the Biami were most closely connected.

We managed to explain that we needed more food and they seemed to say that they would bring some the following morning – which they indicated by pointing at the sun and showing where it would be when they next proposed to come.

So that day, we didn't march. And the following morning, they appeared, just as they said they would, carrying green bananas and taro roots. Cakes of salt were handed over. But no women or children had come with them. Laurie said that was a bad sign. We were not yet friends or sure of one another.

We tried to suggest by signs that we would like to meet more of them and perhaps even their women and children, and they suggested, also by signs, that we should follow them. So leaving all the carriers behind so that we did not appear like an invading army, Laurie, Hugh the cameraman and I did so. They led us into thick forest. And then suddenly – they weren't there.

We called 'Biami! Biami!' but there was no response. Had we walked into a trap? Should we retreat? Or should we go on and risk being ambushed? We decided to go on. We followed their tracks for another few minutes and then we came to their camp. But it was deserted. There was a simple shelter of withered leaves. Beneath it, a small fire, with still smouldering embers. And beside that, on a simple rack of sticks sat a human skull. We called again 'Biami! Biami! ' – but there was no reply. We walked back to camp. And that was the last we saw of them.

So I never did get the chance to follow up my eyebrow lift and smile, or to discover what else remains of the ancient language of gestures that once must have been used by all mankind.

The trust between the Biami and the expedition was far from complete. On the second day of the encounter they fled, leaving behind in their shelter a human skull.

There was little difficulty in bartering with the Biami. They brought food for us and we paid them in a commodity that they greatly valued – salt.

Laurie Bragge, the leader of the Biami patrol used a mobile radio, then a novel and advanced piece of radio equipment, to give instructions for an air drop of supplies.

The peoples of New Guinea, many of them isolated communities in remote jungle-covered valleys, speak about a thousand mutually incomprehensible languages. But trading still goes on between them, so gestures become very important and expressive. One of the Biami shows us how he counts.

Illustrations

Except for those listed below, all illustrations are from the author's personal collection and library. The author is particularly grateful for the generous provision of their pictures by Professor Jeremy Thomas, Dr John van Wyhe, the Natural History Museum of Münster, and SAIAB – the South African Institute for Aquatic Biodiversity.

Index